SHUTTLE 3

NIGEL MACKNIGHT

Motorbooks International
Publishers & Wholesalers ®

NIGEL MACKNIGHT
DESIGN: TERRY GODFREY
FILMS: STEVE PENDLETON

This edition published in 1991 by Motorbooks International Publishers and Wholesalers, P.O. Box 2, 729 Prospect Avenue, Osceola, WI 54020, USA.

Produced by Macknight International

ISBN 0-87938-553-7

Pre-print services:
GA Graphics, Stamford,
Lincs., England.

Printed and bound in Singapore
by PH Productions

Flights courtesy of

CONTENTS

A THOUSAND DREAMS 4
THE EXTRAORDINARY ORBITER 10
LAUNCH! 22
LIFE ABOARD THE SPACE SHUTTLE 50
RETURNING TO EARTH 64
THE ASTRONAUTS 72
THE MAJOR MALFUNCTION 80
BACK TO STRENGTH 88
FINANCIAL YEARS 94
SHUTTLE MENU 95
AT THE CONTROLS 96
INTERVIEW: ASTRONAUT TAMMY JERNIGAN 102
SHUTTLE AS A STEPPING-STONE 106
A VIEW OF THE FUTURE 108

Acknowledgments

NASA/Johnson Space Center, Houston, Texas.
Dr. Aaron Cohen, Rick Hauck, Dave Hilmers, Steve Nagel, 'Hoot' Gibson, Brewster Shaw, Mary Cleave, Kathy Sullivan, Hank Hartsfield, Jerry Ross, Bruce McCandless, Karl Henize, Fred Gregory, Jim Buchli, Bryan O'Connor, Terry White, Steve Nesbitt, Marsha Kracht, Jack Riley, Dick Tuntland, Brian Welsh, Jeff Carr, Kyle Herring, Billie Deason, Barbara Schwartz, Marilyn Davison, Lisa Vazquez, Mike Gentry, Eileen Walsh, Christine Mason, Pat Patnesky, Al Pennington, Tommy Holloway, Larry Bourgeois. Carlos Sanchez, Kari Thornton.

NASA/Ellington Field, Houston, Texas.
Joe Algranti, Ace Beall, Frank Marlowe, A.R. Roy, Kandy Hosea, Hubert Cook.

NASA/Kennedy Space Center, Florida.
Robert Crippen, Ed Harrison, Hugh Harris, Tip Talone, Conrad Nagel, Ann Montgomery, Dick Young, Bruce Buckingham, Lisa Malone, Elliot Kicklighter, Frank Merlino, George Diller, Jo-Ann Mattey, Marion Richardson.

NASA/HQ Washington D.C.
Jim Ball, Mark Hess, Dr. John Lawrence, Les Reinertson, Debby Rahn, Bob Schulman, Charles Redman, Dwain Brown, Dave Garrett, Tom Jaqua, Arnie Aldrich, Jakki Foster, Althea Washington.

NASA/Goddard Space Center, Greenbelt, Maryland.
Dr. Al Boggess, Jim Elliott, Dave Thomas, Dr. Werner M. Neupert.

NASA/Langley Research Center, Hampton, Virginia.
Keith Henry, David Throckmorton.

NASA /Marshall Space Flight Center, Huntsville, Alabama.
Bob Lessels, Ed Medal, Dominic Amatore, Robert K. Ruhl.

NASA/Dryden Flight Research Facility, Edwards, California.
Milton Thompson, Ted Ayers, Lou Steers, Jim Stewart, Ed Schneider, Rogers Smith, Gordon Fullerton, Bill Dana, Steve Ishmael, Tom McMurtry, Don Evans, Laurel Mann, Don Gatlin, Jenny Baer-Riedhart, Dave Lux, Jim Thompson, Ralph Jackson, Nancy Lovato, Pete Waller.

NASA/Caltech Jet Propulsion Laboratory, Pasadena, California.
John Casani (and Lyn), Milt Goldfine, Taylor Wang, Yuri van der Woude, Mary-Beth Murrill, Jim Wood, Jim Doyle.

Boeing Aerosoace, Seattle, Washington.
Donna Mikov, Dr. Dana G. Andrews.

Eaton Corporation, Deer Park, New York.
Dick Dunne, Rich Palmay.

Hughes, El Segundo, California.
Richard Dore, Barbara Bansmer.

Lockheed Missiles and Space Co., Sunnyvale, California.
George Mulhearn, Jan Wrather, Roger Beall, Steve Pehanich.

Martin Marietta, Michoud, New Orleans and Denver, Colorado.
Art Koski, Jan Timmons, Tom Brannigan, Judy Herb, Evan McCollum, Bill Vaile, Doris-Lyn Silby.

McDonnell Douglas Astronautics Co., St. Louis, Missouri and Huntingdon Beach, California.
Susan Flowers, Charlie Walker, Tom Williams, Jim Schlueter.

Morton Thiokol, Birmingham City, Utah.
Rocky Raab, Carver Kennedy, Karen Kerkove, Charles E. Saderholm, Frank W. Call, James E. Brown.

Rockwell Internation, Los Angeles, California.
Bob Howard, Richard E. Barton, Kathy Herrity.

Rocketdyne, Canoga Park, California, USA.
Robert Paster, Joyce Lincoln.

Spacehab, Inc., Seattle, Washington, and Washington D.C.
Robert Citron, Wendy Skony, Jurgen Trohl, Tom Taylor.

TRW Space & Defense, Redondo Beach, California.
Montye Male

Also...
Russ Brown (illustration on page 11), Bob and Kit Overmyer, Peter Joel of Pan American (travel arrangements), Jim Lovelady of the Public Affairs Department of the U.S. Army at White Sands, New Mexico, Klaus-Peter Wenzell of ESA/ESTEC, Noordwijk, Holland, and June Scobee, Glennis Yeager, John Manke, Fitzhugh Fulton, Bill Andrews, Trieve Tanner, Joe Engle, Byron Lichtenberg, George Economy, Vinit Nijhawan, Robert L. Stewart, Joe Cupido, Gaz Top.

FOREWORD

On 14 October 1947, Brig. Gen. Charles 'Chuck' Yeager – flying the Bell X-1 research aircraft – became the first man to travel faster than the speed of sound.

That day comes back to me every now and then. We were up around 0.96 Mach number and we were in quite a bit of buffeting. The airplane just shook. Then, all at once, the Mach-meter just jumped from off the scale, and at this point all the buffeting quit and the airplane smoothed out just as slick as glass...

Those early days with the X-1 showed us how much could be achieved when a small but dedicated team of people get together to move one step further than the others.

NASA's Space Shuttle team – and the teams from the contractor companies, large and small – are making tremendous progress in their efforts to put space to work. The benefits of this effort will be felt by almost everyone, be it through improved telecommunications, the orbital monitoring of Earthly problems such as crop disease, or just the technological 'spin-off' which brought us the silicon chip, advanced construction materials such as carbon fiber, and fiber optics.

Nigel Macknight's fine book sets out the Space Shuttle story in a no-nonsense manner and is beautifully illustrated. I shall be well pleased if those who read it feel inclined to further their interest in this infinite field of human endeavor.

Chuck Yeager

The Author

NIGEL MACKNIGHT was born on 8 August 1955 in the Roman town of Corbridge, Northumberland. He developed an early passion for aviation – the result of being brought up near the English Electric (later British Aircraft Corporation) facility at Samlesbury, Lancashire, where twice-supersonic Lightning interceptors flew daily.

At college in Nottingham, Macknight studied art and developed his part-time activities contributing to various aircraft magazines: a pastime he began in Glasgow in 1971. In 1977, he took up writing on a full-time basis, and wrote articles and authored and produced books on aviation and other subjects for various publishers until 1983/84, when he became a publisher in his own right and began making occasional forays into television and radio broadcasting.

The original edition of Shuttle *was Macknight's first offering as a publisher. After enjoying several updates and reprints over the years, sales are already in excess of 40,000 and should exceed 50,000 with the release of* Shuttle 3. *The* Shuttle *book also appeared in magazine form, as* Shuttle Story: *a seven-element partwork published between April 1984 and August 1985. That gave rise to an all-new monthly magazine,* Spaceflight News, *which Macknight edited and co-published in conjunction with Key Publishing of Stamford, Lincolnshire from October 1985 to November 1990, then operated ndependently for several months thereafter.*

Sadly, Spaceflight News *then ceased publication, having run for a total of 62 issues. Nigel Macknight, though disappointed by* SFN's *closure, has turned once more to authoring books on aerospace topics. He now hopes to establish his* Space Year *annual*

Author's Note

IN THE panel at left, I've tried to acknowledge everyone who helped bring this book to fruition in its various editions, including its expression in magazine form. If I've missed anyone out, I'm sorry.

There are two people I'd particularly like to thank at the start of this, the third updated edition. One is Terry Godfrey, who made outstanding contributions in both time and skill to help me produce the original book, long before there was any guarantee that the design abilities he brought to it would ever see the light of day.

The other is David Henley, Managing Director of the Beatties hobby-shop chain. David also saw the book's potential and assisted its progress to publication in a practical manner. I'd like to think that, as sales approach the 50,000 mark, he takes some satisfaction in having had that initial faith.

A THOUSAND DREAMS

A study in progress. Way back in 1963 the diminutive M2-F1 — secretly tested to ascertain the airworthiness of the radical wingless concept before NASA headquarters was informed — helped pioneer the re-entry flying techniques which led to the Shuttle Orbiter.

Space Shuttle is a revolutionary craft, make no mistake about that. But a clear ancestry can be traced back from it. Numerous designs — some of which were actually built and flown, while most never even left the drawing board — made their contributions to the fund of knowledge from which the Shuttle sprang forth. When the American newspaper *Today* described the Shuttle as the "Son of fifty fathers, a thousand dreams", they weren't far wrong.

The impetus to create a reusable spacecraft came from economic pressures. Far too much expensive 'garbage' was being left to float in free orbit around the Earth after use, eventually to tumble back towards a fiery destruction on contact with the denser levels of the atmosphere. On top of that, the cost of recovering manned capsules from the ocean once they had been parachuted to a safe splashdown was astronomical. But above all it was the wastage of all the sophisticated rocketry which put that manned capsule into orbit in the first place that was becoming increasingly unjustifiable. Stage by stage, the booster segments fell away during ascent as their fuel was expended — all to provide orbital insertion for the one tiny component that had sat atop the great white leviathan only minutes before on the launch pad.

In their efforts to retain the fuel-carrying component for future reuse, NASA considered all manner of unconventional designs. Perhaps the most spectacular of these was the Hyperion proposal, which envisaged the retention of the propellant-carrying vehicle on a massive railway line running horizontally for 3.2 kilometres across a stretch of desert, then curving up a near-vertical mountainside for another 1.6 kilometres. The capsule-like Hyperion would guzzle fuel from the propellant

A dramatic view of the M2-F2 manned lifting body employing rocket propulsion to blast its way up to higher speeds and altitudes prior to initiating a carefully controlled gliding descent to a pinpoint landing at Dryden Flight Research Center, Edwards, California.

The Boeing X-20 Dyna-Soar — cancelled in 1963 before it could be flown — was a notable precursor of the Shuttle Orbiter.

vehicle to power its cluster of rocket engines almost to the top of the mountain, before parting company at a speed of over 1,000km/hr to make its own way up into orbit using a relatively small amount of internal fuel. On separation, retrorockets could bring the propellant 'sled' to a halt and send it back down to its base for refuelling and reuse when necessary. If built, the propellant 'sled' would have been by far the largest device ever to run on rails, but the Hyperion vehicle itself was no babe-in-arms either. It was intended to carry up to 110 passengers on a single trip! Re-entry would be achieved by the method employed by the conventional space capsules of the time, but the usual parachutes would be replaced with a battery of retrorockets which would lower the craft gently to earth.

Of course, this project was just a bit too ambitious to succeed. But the fact that it received serious study gives some idea of the lengths to which NASA's design teams were willing to go in those early days to solve the reusability problem.

The total value of hardware abandoned in space during the Apollo missions has been estimated at $24 billion. In the early days, the extravagant 'throwaway stages' method was justified by the politicians whose job it was to approve the huge financial expenditure it entailed. It was seen, perhaps rightly, as the price that had to be paid for beating the Soviet Union to the surface of the Moon. Developing a reusable system would take ten years or more — and the flag bearing a Hammer and Sickle would be firmly ensconsed in the lunar dust by then for all to see.

The ultimate answer to the wastage of vastly expensive space hardware is the so-called 'single stage to orbit' concept; a spacecraft which carries everything it needs to get into orbit, do its work there and return safely, without the need to jettison anything. Even in the days when what is now the Space Shuttle was beginning to take shape on numerous sketchpads across the USA, this all-capable concept was recognised in certain quarters as being achievable. But it would not be easy — and it would not be cheap. As economical as it might have been once in service, its initial research and development cost would make all previous space budgets pale to loose-change proportions by comparison, so the emphasis quickly changed to that of a reusable system comprising of a number of integrated, but jettisonable, components. Some earlier ideas and ventures provided food for thought.

Way back in 1947, an edition of the *Sacramento Bee* newspaper dated 13th February carried the first part of a serialised feature on a proposed reusable rocket ship which could be launched from the belly of a large airliner to make repeated flights to the Moon. Beneath an artist's impression depicting the moment of launch, and a cutaway diagram of the rocket ship itself, were the words, 'All aboard for the Moon! The rocket ship will leave at midnight. Back to Earth for late breakfast. Round tickets only'. Although this bizarre spacecraft was based on the 'single stage to orbit' concept, it is interesting to note that some of its technical features appear on the Space Shuttle of today. For example, for manouvering purposes there were to have been four small jets in the stern set at right angles to one another. By means of these four jets, the spaceship could be turned in any direction, and even twisted about any axis at will. Like today's Space Shuttle, it could travel just as easily stern-first as it could travelling nose-first. The OMS, or Orbital Manouvering System, which appears on the Shuttle Orbiter performs just such a task over three decades later.

Much later, a series of more plausible design concepts helped pave the way for the routine use of space. One such concept gained much publicity during its short-lived period in favour in the early 'sixties. This was the Boeing X-20 Dyna-Soar, a diminutive sub-orbital aircraft with stubby delta wings which curved sharply up to form a vertical finlet at each wingtip. This general concept had first been formulated by a German, Doctor Eugene Saenger, in the book *The Technique of Rocket Flight* published in 1933. By the summer of 1942 German research into his ideas had been abandoned, ironically for the same reason

A Shuttle forerunner, the Northrop-built HL-10 lifting body (HL stands for horizontal landing) used its body shape to generate aerodynamic lift for flight, eliminating the need for wings.

North American Aviation built the rocket-propelled X-15 hypersonic research aircraft, jointly operated by NASA and the U.S. Air Force. It eventually reached a maximum altitude of 354,200 feet and an astonishing maximum speed of 4,543mph.

the Americans later turned their backs on winged, reusable rocket craft of the X-15 *genre* in the 'sixties; the Germans considered it more expedient to concentrate on straightforward ballistic designs like the V-2, just as the Americans later favoured the "spam in the can" Apollo-type method of getting to and from space.

The name Dyna-Soar was an abbreviation of the vehicle's proposed DYNAmic take-off and SOARing, or gliding, re-entry capabilities. It would be launched atop a huge Martin Titan III booster and return to land on a conventional runway, having first jettisoned a protective outer cladding of heat-resistant material designed to withstand the extreme friction that would build up between the vehicle's skin and the increasingly dense levels of air which would surround it on re-entry. Sadly, Dyna-Soar did not get beyond the full scale mockup stage before President John F. Kennedy made his famous announcement that the U.S. space effort should be wholeheartedly directed at getting man onto the Moon. At this time the X-20 was already the subject of an unrelenting attack by the-then Secretary of Defense, Robert McNamara, who slated the Dyna-Soar concept as grossly uneconomical.

Although the project was cancelled in December 1963, the Dyna-Soar provided a pointer of things to come. No less than seven aerospace companies had tendered proposals in response to the USAF's initial requirements, producing a broad base of both theoretical and wind-tunnel test data which would prove useful to future ventures. As well as being a notable precurser of the Space Shuttle Orbiter concept, the Dyna-Soar had many other features which were highly advanced for its time. These included a side-stick control column similar to the type now used on the F-16 fighter aircraft, a 'puffer-jet' control system for manouvering in the airless regions beyond the atmosphere, and an advanced monitoring computer which operated a 'fly-by-wire' flight control system. A system of this type dispenses with long runs of activation cable and push-pull control rods to the various control surfaces and replaces them with wiring which carries command signals direct from the pilot to servos on the control surfaces. This results in a significant weight saving and reduced complexity.

Dyna-Soar never flew. But other Shuttle ancestors did. Where does conventional flying end and spaceflight begin? The rocket-propelled Bell X-1 which carried test pilot Chuck Yeager through the notorious 'sound barrier' in October 1947 flew to high altitudes and — like the Shuttle Orbiter — glided earthwards onto a conventional undercarriage. But for all its magnificent achievements, the X-1 was still only an aeroplane pointing the *way* to space. The first winged machine truly to penetrate the fringes of space itself was the North American X-15 — one of Space Shuttle's most significant ancestors.

The immediate aims of the X-15 project were to gather data on aerodynamic behaviour at hypersonic speeds, develop reaction controls to enable the vehicle to manouver in the airless regions high above the earth (shades of the 'puffer-jet' control system planned for the Dyna-Soar), verify the ability of such a craft to re-enter the atmosphere without burning up, and to develop a means of manouvering in the upper atmosphere in such a way as to enable a pre-selected landing area to be reached with great accuracy. NASA knew that the achievement of these objectives would be vital to the development of a future reusable space vehicle, and two complimentary programmes helped bring that to fruition. One was the USAF's START (Spacecraft Technology and Advanced Re-entry Test), initiated in 1961, while an advanced phase of this programme was called PRIME (Precision Recovery Including Manouvering Re-entry).

The X-15 effectively bridged the gap between the conventional aircraft then in service and the early Mercury and Gemini capsules which pioneered America's manned space programme. Operated jointly by NASA and the U.S. Air Force, it was a needle-like aircraft with tiny stub wings and a wedge-shaped tailfin, the better to deal with the very high temperatures involved. A great deal had yet to be learned about high temperature airframe operation. Accordingly, the X-15's skin was constructed from a nickle-based alloy called Inconel X, while much of the interior structure was titanium alloy. Fusion welding techniques were widely used during manufacture. With propulsion coming from a single Thiokol XLR-99 rocket motor developing 57,600 pounds static thrust, the X-15 eventually reached an altitude of 354,200 feet (107,960 metres) and an astonishing maximum speed of Mach 6.72, or 4,543mph (7,297km/hr).

The X-15 did not take off from a runway like conventional aircraft. Instead, it was carried aloft under the starboard wing of a modified Boeing B-52 bomber to conserve sufficient fuel for its research flights. Once it had been carried to a high enough altitude, the X-15 was literally dropped like a stone. Its XLR-99 rocket engine then sprang into life, and it simply disappeared into the distance in a matter of seconds. Travelling at phenomenal speeds and altitudes the X-15 performed 360-degree rolls and other manouvers to evaluate hypersonic handling characteristics. In fact, the altitudes attained enabled some of the X15's pilots (which included future Moon-walker Neil Armstrong) to be officially termed 'astronauts'.

Once powered testing began (following a series of glide-tests commenced in mid-1959) the X-15's speeds increased inexorably. And so did the searing temperatures on its airframe. The highest temperature actually recorded during the rocketplane's career was 1,323

Among the intriguing ideas studied in the early days was a proposal to use a ship to move the Orbiter from place to place, instead of a 747 aircraft.

This picture of the 'wooden' M2-F1 (left) and the rocket-propelled, metal M2-F2 lifting bodies illustrates the evolutionary nature of NASA's wingless aircraft programme of the late 'sixties/early 'seventies. The nose windscreen aided pilot orientation during landing.

degrees F (716 degrees C), as opposed to the expected figures of 800-1,200 degrees F (430-650 degrees C). Herein lay the value of the X-15 test programme, for inaccuracies in both theoretical calculations and predictions based on windtunnel tests were shown up.

Flight testing of the X-15 taught the aerodynamicists a great deal. Firstly, it became clear that discontinuities in the aircraft's skin should be avoided, because these gave rise to local stagnation temperatures. Any orifice which allowed the hot layer of air surrounding the aircraft's exterior surfaces to get through to the interior had to be sealed. An uncovered slot in the wing leading edge, for example, produced a 'hot spot' and consequent buckling of the aircraft's skin. A leak through the nosewheel door burned through aluminium tubing and sent smoke into the cockpit, while others damaged instrumentation wires at the tailplane roots and gave rise to excessive temperatures around the air-brake actuators.

Another lesson learned related to the incorporation of expansion joints in those areas where an extremely hot structure met an extremely cold one. One such area was the lateral fairings, where the temperature gradient between the super-hot 600 degree F (315 degree C) and the super-cool −260 degree F (−162 degree C) liquid oxygen tank caused buckling of the secondary structure. Similar problems were encountered during early test-firings of the Space Shuttle's main engines many years later. Some trouble was also experienced with the X-15 when a windscreen shattered. Although the aircraft's alumino-silicate glass had been theoretically cleared to 1,000 degrees F (538 degrees C), this damage was caused as a result of the windscreen frame buckling and causing a local hot-spot.

The second of three X-15As built landed without flaps on 9 November 1962, collapsed its undercarriage and ended up on its back. During repairs, the opportunity was taken to modify this aircraft to permit even higher speeds to be attained, making it suitable for trials of externally-mounted ramjet engines at up to Mach 8. (NASA and North American stated that Mach 9 could be achieveable with an uprated XLR-99). In order to increase rocket burning time from 83 to 150 seconds, two enormous drop tanks were fitted to X-15A-2's flanks, just beneath the wing roots. These would be jettisoned 65 seconds after launch; a procedure later successfully demonstrated. Sadly, although the X-15 flew in this guise, it had never been taken anywhere near its limits by the time the test programme was terminated in November 1968 after 199 flights had been made.

If the advanced configuration had flown, however, far higher structural temperatures would have been encountered. These were estimated to reach 2,400 degrees F (1,315 degrees C) along the leading edges of the wings, thus exceeding the 1,200 degrees F (650 degrees C) limit of Inconel X over much of the skin area. For this reason it was proposed that approximately 300 pounds (137 kilograms) of T-500 ablative material be applied to the Inconel X by spray-gun in thickness of up to 0.7 inch (1.78cm). Ablative material disperses heat by shedding glowing bits of a heavy, resinous substance and was used on the Mercury, Gemini and Apollo re-entry capsules with great success. Alternatively, even more advanced thermal protection materials could have been installed; columbium wing leading edges, and skins of Rene 41 — an 1,800 degree F (980 degree C) nickle-base alloy originally slated for the Dyna-Soar.

Another advanced-concept plan for the aircraft involved converting X-15A-3 to a delta wing configuration. In retrospect, this would have provided a fascinating insight into the basic flight characteristics of the Space Shuttle, which utilises a similar wing shape. Alas, it was not to be. . .

Many years later, the Commander on the first Space Shuttle mission, John Young, recognised the X-15's important contribution to the reusable space vehicle concept when he said, "The lift-to-drag ratio of the Space Shuttle is almost identical to that of the X-15. They were very similar programs and there was a great deal of feedback from the X-15 into the Shuttle. It really paid off". The man at John Young's side on that momentous maiden flight — Robert Crippen — agreed; "The experience that we gained from an aerodynamic standpoint, energy standpoint and basic

The X-24 lifting body is launche from a B-52 'mother ship' flying at 45,000 feet. Under rocket power it climbed to higher speeds and altitudes before manoeuvering to a glide landing on a dry lake bed at Edwards, California.

The Northrop HL-10 lifting body — with its astronaut pilot nearby — is overflown by the Boeing B-52 which only minutes before had launched it on a vital test sortie. Note the sun-parched dry lake bed runway surface.

rocket technology was the kind of thing that made the Space Shuttle possible".

X-15 wasn't the only aircraft built to explore re-entry flight performance. Studies and tests of a concept known as the 'manned lifting body' began in the USA in the early 'fifties, and by the time the X-15 had made its first glide-test, NASA were considering introducing yet more revolutionary shapes to the skies. In principle, the aims of the lifting body programme were not that far removed from those of the X-15, though particular emphasis was placed on increasing the degree of manouverability that could be achieved during re-entry. But this time, research would focus on proving the feasibility of *wingless* craft which obtained their aerodynamic lift solely from the shapes of their bodies. The immediate results were the extraordinary vehicles designated M2-F2 and HL-10.

The reasoning behind the elimination of wings was simple; at a projected re-entry speed of Mach 25, wings were deemed unnecessary for the atmosphere to support them, and by the time a craft entered the denser levels of air it would still have enough speed to rely on its fuselage shape for such lift as it required to maintain a steep descent path. On top of that, there was at this time no practical solution in sight to the problem of dissipating the extreme temperatures of re-entry with a low-weight thermal protection system. Ablative materials were just too heavy. By dispensing with wings — thin edges and appendages generate the most heat — this problem could be cleverly avoided. The manned lifting bodies were based on a conical shape, because of its ability to remain stable during hypersonic flight (the re-entry capsules of Mercury, Gemini and Apollo were all conical). NASA's Ames Research Center in California calculated that a craft based on a 13 degree cone, but with its upper surface flattened off, would provide the optimum basis for a manned vehicle. The resulting M2 lifting body differed from the basic configuration only in the adoption of a rounded snout to alleviate local overheating. Meanwhile, NASA's Langley Research Center in Virginia came up with the HL-10, which differed from the rival Ames configuration in being flat-bottomed rather than flat-topped. Both the M2 and the HL-10 were built by the Norair Division of the Northrop Corporation. They both glided to a landing, but powered themselves to high altitude by means of their own rocket engines after being air-launched by the trusty Boeing B-52. In addition, they were fitted with tiny throttlable rockets to help smooth out the final moments of a hair-raising thirty-degree landing approach. This was about ten times steeper than a conventional aircraft's angle of descent...

The Ames proposal was first tested in a form designated M2-F1 in 1963; the same year that the Dyna-Soar project was cancelled. The F1 variant was a low-cost experimental model with an internal tubular structure, an external shell made almost entirely of plywood, and a fixed undercarriage. It was completed in a matter of months by a sailplane manufacturer. Two vertical fins and rudders controlled yaw, two trailing edge flaps controlled pitch, and two differential elevons controlled both roll and pitch simultaneously. The M2-F1 was tested in Ames' full-scale windtunnel, and successful practice flights were conducted by towing it at low level behind a powerful car at speeds up to 125mph (200km/hr) before a proper air-test programme could begin. Towed behind a Douglas C-47, the M2-F1 glider proved the feasibility of the lifting body concept by making over 100 flights in the hands of experienced test pilots.

It is a little-known fact that construction of the M2-F1 took place without NASA headquarters' knowledge. Ames Center Director, Paul Bikle, took it upon himself to build it and make the first few flights without anyone in authority knowing about it. Only when the initial proving flights had shown that the radical wingless concept was basically airworthy was NASA headquarters informed...

The lightweight F1 model was followed by the rocket-propelled F2 derivative, which was rolled out on 15 June 1965. The M2-F2 had a retractable undercarriage and other refinements, and propulsion came from a single Thiokol XLR-11 alcohol-liquid oxygen four-chamber rocket motor generating 3,630 kilograms (8,000 pounds) static thrust for a 100-second 'burn'. The M2-F2's maiden flight from under the B-52's starboard wing was conducted by NASA's Milton O. Thompson on the morning of 12 July 1966. It was dropped from an altitude of 13,700 metres (44,936 feet) and landed 3½ minutes later.

The Langley-inspired HL-10 made its maiden flight some time after the M2-F2 had done so. In fact, the M2-F2 had reached flight fourteen in its test programme by this stage. Bruce Peterson flew the first HL-10 mission on 22 December 1966, but his next flight — the sixteenth M2-F2 flight — was a disaster. Whilst approaching the landing site after an unpowered approach, the lifting body began to oscillate violently. Peterson altered the HL-10's flight path to help compensate, then did so again to avoid the helicopter which was flying nearby to assist the timing of the landing flare out. With the undercarriage only partly down he attempted a landing on featureless desert terrain nearby. A notoriously difficult place to judge distance and altitude in an already critical situation. The M2-F2 smote the sand at 217mph (350km/hr), and rolled six times before it came to a standstill and lay in sickening silence. Amazingly, Bruce Peterson survived this accident, though his aircraft was very badly damaged. Peterson's miraculous recovery in the months that followed provided the inspiration for the popular television series, *Six Million Dollar Man,* which shows actual film of the terrifying desert accident at the beginning of each episode.

The M2-F2 was rebuilt from the ground up by 1969 — redesignated the M2-F3 — and featured reaction controls similar to spacecraft thrusters, and the addition of a third vertical fin assembly. This aircraft was retired in December 1972, having reached a maximum altitude of 71,504 feet (21,800 meters), and a maximum speed of Mach 1.613.

All manned lifting body flights up to this point had been unpowered, but Major Jerauld R. Gentry made the first rocket-powered flight (in the HL-10) on 23 October 1968. At around this time, the lifting body programme became jointly sponsored by the USAF and NASA, and a third aircraft type — the Martin Marietta X-24A — joined the test programme. Developed from the

The X-24B (first flown 1971) was an advanced-concept development of the original X-24A. Its shape came closer than the other lifting bodies to that of the Orbiter.

unmanned X-23A (itself a derivative of the SV-5 series of re-entry glider scale models) the X-24A featured a flat-topped fuselage, a glazed nosecap to aid forward visibility, three tailfins and a Thiokol XLR-11 engine. Its first gliding flight was made on 17 April 1969 with Jerauld Gentry at the controls, followed by a series of powered flights culminating with its first supersonic flight in October of the following year. Its highest speed was Mach 1.60, achieved on June 4 1971, whereupon it was extensively modified to X-24B standard. With a considerable increase in nose length, an elegantly curvaceous fuselage, and increased all-up weight and width, the X-24's shape came closer than any of the other manned lifting bodies to that of the Space Shuttle Orbiter. Over 100 lifting body flights (excluding those flown by the 'wooden' M2-F1) had put America's space programme firmly on course for success in its forthcoming quest for vehicle reusability.

While strange shapes took to the air and glided back to earth, frenetic activity had also been taking place on drawing boards right across the USA. The Space Shuttle as we know it today was slowly beginning to emerge. One recurring theme for the new reusable spacecraft involved a pick-a-back layout, with two machines taking off vertically locked firmly together. One would provide the fuel and thrust for the initial climb away from the launch pad, before separating to allow the prime vehicle to continue the journey into orbit under its own power once the 'mother ship' was spent. This latter vehicle — with a two-man crew in command — would then glide back to a conventional landing, to be followed by the orbital craft once its mission in space was completed. The orbital craft would also have a two-man crew, but would provide additional accommodation for twelve payload specialists.

Some of these pick-a-back designs studied by NASA involved ultra-sleek vehicles with gracefully curving wings and dolphin-like fuselages, while others had a shape not that far removed from that of today's Orbiter; a chunky, almost amphibious-looking design with a curved 'hull' to carve a path through the upper atmosphere on re-entry. Either way, it was intended that the combination would be propelled by two high-pressure hydrogen/oxygen rocket engines; two or three for the Orbiter and ten or twelve for the booster vehicle. By making these engines throttlable to half power during ascent, G-force levels could be reduced to the point where ordinary people could be carried rather than Supermen, and payloads would not be subjected to such heavy punishment. At a slightly later stage, payload bay doors set into the upper fuselage began to appear on the draughtmans' models and drawings. A cargo bay would have to be big enough to ferry components for a space station or similarly large structures from Earth into orbit, or deploy and collect weather and communications satellites.

The pick-a-back concept looked quite promising both on paper and in the wind tunnel, but the cost estimates told a different story. At 1970 prices the project would have set American taxpayers back to the tune of at least $10 billion. The project assessment chief at Kennedy Space Center, Sam Beddingfield, decided it was time to "throttle back" to a more realistic proposal.

After well over one hundred different design proposals had been considered, that more realistic proposal turned out to be today's four-part Space Shuttle vehicle, financed to the tune of $5 billion by federal budgetmakers and the-then President, Richard Nixon. Inflation has since pushed that figure up above the $10 billion mark. The four primary components of today's Space Transportation System (STS) are the manned Orbiter vehicle, the giant external tank which supplies liquid hydrogen and liquid oxygen fuel to the Orbiter's three main engines and provides a structural backbone for the Shuttle vehicle in its launch configuration, and — last but by no means least — two solid rocket boosters, which provide the 'kick' necessary to get the heavyweight craft up off the launch pad and on through the denser levels of the atmosphere. Other, secondary components in STS include the Spacelab orbiting laboratory, a gigantic Space Telescope, a system of tracking and data relay satellites (TRDSs) to aid navigation and transmit information on payload operations, and extensive checkout, launch and landing facilities. But these are the periphery facilities. Spiritually, and certainly in the eyes of the general public, the 'Shuttle proper' is the reusable manned Orbiter — last in a line of truly extraordinary ancestors; "Son of fifty fathers, a thousand dreams"...

This artist's impression shows a typical 'pick-a-back' configuration spacecraft of the type considered before the familiar Shuttle shape of today was finalised in the early 'seventies. The lower machine would have carried most of the ascent propellant.

THE EXTRAORDINARY ORBITER

A Shuttle Orbiter under construction at Rockwell International's Palmdale, California plant. Most of the airframe is fabricated in a conventional manner from aluminium alloy panels, frames and bulkheads. The tail is built by Fairchild at Farmingdale, New York.

Most people who see the Space Shuttle out on the launch pad agree that its most exciting element is the Orbiter vehicle. This may be due in part to its fascinating shape, but mainly it's because the chubby spacecraft represents the *human* aspect of this dauntingly technological venure. There cannot be many people who view the white arrowhead-shaped craft at close quarters without gazing up in awe at the cockpit area and feeling a surge of admiration for the men who will soon be ensconsed within it and hurtling up into the black void of space.

The Shuttle Orbiter first took to the air after being air-launched from a specially-converted Boeing 747.

The 747 'mother' plane is officially titled the Shuttle Carrier Aircraft, or SCA. It continues to play a vital role in the STS programme, having initially been used to carry the Orbiter *Enterprise* aloft for the series of Approach and Landing Test (ALT) flights conducted over Dryden Flight Research Center — situated 'next door' to Edwards AFB — during the course of 1977. These tests enabled NASA to gather data on the spacecraft's gliding performance, and to confirm its ability to make a safe 'deadstick' landing of the type that would later follow re-entry into the Earth's atmosphere.

In the years following the ALT flights, the SCA 747 has been used to ferry Orbiters from one place to another; a task that it has performed with faultless reliability. Indeed the SCA has become something of a celebrity in its own right, catching some reflected glory from the Orbiter's outstanding success. But the story of its conversion from a standard aeroplane is not so well known.

The Boeing 747 airliner was but one of several potential Shuttle-shifters considered for the job. Other candidates included the C-5A Galaxy and a totally new aircraft configured to carry the Orbiter beneath a stub wing linking two fuselage sections positioned side-by-side. At one point it was even considered feasible to fit turbofans to the Orbiter to enable it to get around under its own propulsion, but this idea was eventually dropped when it was calculated that the spacecraft generated insufficient lift to provide the required range with the amount of fuel it could carry.

The 747 registered N905NA started its operational career back in October 1970 when it was purchased from the manufacturers by American Airlines. After accumulating 8,899 flying hours and 2,985 landings on regular scheduled operations it was purchased by NASA in July 1974 for $15 million, and was put to use in a flight research programme conducted by the space agency's Dryden facility to investigate the problems associated with wake vortex flow from widebodied jet transports. On completion of this programme, the 747 was returned to Boeing's Seattle, Washington plant in April 1976 for the modifications which would convert it into the SCA.

A comprehensive series of modifications was necessary to alter this standard 747-100 model for its unique new role. Its main structure was reinforced to support the Orbiter's 150,000-pound (67,500-kilogram) weight, and three supports — one forward, two aft — were positioned atop the fuselage to locate onto attachment points on the spacecraft's underbelly. These attachment points would later be used to connect the Orbiter to its giant external tank prior to a space mission. The Orbiter would require two different angles of inclination when atop the SCA,

A Rockwell International technician affixing black thermal protection tiles to the area surrounding the aft Reaction Control System nozzles.

depending on the type of flight involved. A slight nose-up angle would be needed for the ALT sorties, while ferry flights would dictate an angle of inclination more or less parallel to the SCA's line of flight. Two vertical fins braced by streamlined struts were fitted to the tailplane extremities to aid directional stability when the Orbiter's presence atop the SCA partially 'masks' the ex-airliner's main fin/rudder assembly from the airflow.

The application of NASA's neat red, white and blue livery completed the external transformation — though the American Airlines title logo can still be seen from certain angles when the sun catches the aircraft's sweeping silver flanks. But, under the skin, still further modification had taken place. The 747's airline-specification galleys and most of the passenger seats were removed to save weight, and modifications were made to the cockpit, provision being made for the controls and displays necessary for mid-air launches and piggyback ferry flights. A slide-escape system was also provided, so that the Boeing's NASA crew could 'abandon ship' if things went badly awry. An additional modification took the form of a mass inertia damper installed in the 747's forward fuselage. This apparatus consists of a 450-kilogram mass that shifts position laterally by means of rollers set into the cabin floor, thus damping out the oscillations caused by air flowing over the Orbiter and inducing turbulence across the Boeing's tail.

Other major modifications included the conversion of the aircraft's four Pratt & Whitney JT9D-3A turbofan engines to 7J9D-7AH specification, conferring a useful increase in take-off thrust. With a total height of 75 feet (23 metres) the SCA/Orbiter combination soon became known as "the world's largest biplane".

The ALT flights conducted at Dryden with *Enterprise* mounted atop the SCA were highly successful. On paper, the whole operation looked somewhat precarious — critics of the Shuttle programme had described the Orbiter as a "turkey" that would never fly — but NASA had gone to great lengths to ensure that this would be more than just a "let's try it and see" exercise. On the early test flights *Enterprise* remained mated to the SCA, but later it proved capable of separating cleanly from the 'mother ship' and gliding down to a controlled landing.

Because the ALT flights took place entirely within the Earth's atmosphere, it was not necessary to equip *Enterprise* for space flight. In this respect, she was always a non-standard machine, but she did all that was asked of her. Unlike the operational Orbiters which were to follow, *Enterprise* sported a long air data probe protruding from the nose section, and only a relatively small area of her skin was covered with genuine thermal protection tiles, the remainder being simulated by foam insulation. In addition, the 'engines' protruding from the aft fuselage section were only dummies intended to reproduce the mass and geometry of real SSMEs. Ejection seats were installed for the two-man astronaut crew to use in the event of an emergency.

For the early flights, a seventeen-piece streamlined tail cone assembly built by Boeing and known as the 'ferry pod' was fitted to the aft end of the spacecraft to minimise the amount of aerodynamic

Orbiter *Enterprise* begins its hair raising descent from the tall MVGVT tower at NASA's Marshall Space Flight Center, Alabama, following a series of vibration tests in 1978.

buffeting endured by the carrier aircraft. As its name suggests, this tail cone continues to be used to this day for ferry flights, which are usually flown at a cruising altitude of about 13,000 feet and a speed of 315-342 knots.

The highly-experienced astronauts selected to fly the ALT missions were split into two crews; Fred Haise/Gordon Fullerton and Joe Engle/Dick Truly. The men flying the SCA were no less experienced in their own field. One was Fitzhugh Fulton Jnr., a veteran pilot who had already served as launch pilot of the Boeing B-52 carrier aircraft used to air-launch the rocket-propelled X-15 research aircraft and the wingless manned lifting bodies that had helped to evolve the Orbiter's final shape. Fulton had also put in many flying hours as a project pilot of the radical XB-70 Valkyrie and YF-12A 'Blackbird' aircraft. The other SCA pilot was Thomas McMurtry, who had been flying experimental aircraft for NASA since 1967, including the X-24B lifting body, F-8 Crusader and F-111 super-critical wing and digital 'fly-by-wire' aircraft.

SCA flight test engineers were Louis Guidry Jnr. and Victor Horton. Guidry had previously flown in a similar capacity aboard a C-135 on zero-gravity studies (the aircraft is flown on a parabolic trajectory to create a weightless environment for short periods of time) and in a C-130 Hercules earth-resources studies aircraft, while Horton was formerly a flight test engineer on the YF-12A development programme at Dryden and a launch control panel operator on NASA's B-52 air-launcher. The astronaut crews aboard *Enterprise* could not have wished for a more experienced and enthusiastic team to be flying beneath them into the uncharted territory that lay ahead.

Once taxiing tests at incrementally greater speeds had been successfully concluded with the Orbiter and SCA in their mated configuration, the ALT programme proceeded through three distinct phases;-

MATED INERT TESTS: *No crew aboard Orbiter, no separation.*

For these flights, the Orbiter vehicle was completely inert. That is to say, no crew were aboard and none of its systems were powered up. The aim was merely to assess the Orbiter's structural integrity in the airflow, check out the low-speed handling performance and general handling of the SCA/Orbiter configuration, and simulate the gentle diving manoeuver that would later precede actual separation of the two machines. That the huge streamlined tail cone assembly fitted to *Enterprise* was performing well can be gauged by pilot Fulton's gleeful comment after the first Mated Inert Test; "I couldn't even tell the Orbiter was aboard!"

Although six flights had been planned for this phase of the ALT programme, project manager Deke Slayton — one of the original seven-man Mercury astronaut team — decided that the configuration had flown sufficiently well to render the final flight unnecessary. The entire Mated Inert Test phase had been completed in just twelve days.

Under the skin. This picture of *Challenger* under construction at Rockwell's Palmdale facility reveals the gold mylar blanketing material which is used for insulation.

MATED ACTIVE TESTS: *Orbiter manned, no separation.*

Here, the aim was to prepare the Orbiter for free flight by operating all required systems, get the flight and ground crew up to a level of full readiness, and evaluate the flight conditions that would prevail at the precise moment of separation.

The first flight of this second ALT phase took place on 18 June 1977 with Fred Haise and Gordon Fullerton at the Orbiter's controls. Haise had qualified as a NASA astronaut in 1966, and had been into space on one occasion, namely as lunar module pilot on the ill-fated Apollo ("Houston, we have a problem") 13 mission in April 1970, when an oxygen tank explosion threatened to leave the three-man crew stranded in space. In addition, Haise had served as backup lunar module pilot for the Apollo 8 and Apollo 11 missions, and backup Commander for Apollo 16. Backup crew-members are the spaceflight equivalent of theatrical understudies; fully trained and ready to step in at a moment's notice if one of the prime crewmembers is unable to fly.

Bald-headed Fullerton, on the other hand, had yet to fly in space. Having been a member of the USAF's aborted Manned Orbiting Laboratory programme in the late 'sixties, he joined NASA when the project folded in 1969 and became a member of the support crews for Apollo 14 and Apollo 17.

Although one of the Orbiter's computers had been replaced less than twenty-four hours before take-off, the first 'powered-up' flight — duration 55 minutes — went exactly according to plan.

On the second flight, astronauts Joe Engle and Dick Truly took their turn in the Orbiter's cockpit. Engle, whose energetic sense of humour is well-known and well-loved throughout NASA, looked very calm and collected. Sixteen flights in that fearsome device known as the X-15 during his time as a USAF test pilot at Dryden had provided him with plenty of experience to fall back on. He had also been backup lunar module pilot for the Apollo 14 mission. To Engle's right sat Dick Truly — another ex-MOL flight crew member to join NASA in 1969. Truly had gone on to serve on the support crews for the Skylab flights and the joint United States/Soviet Union Apollo-Soyuz Test Mission in 1975.

Once again, *Enterprise* behaved impeccably — but these were still early days. Throughout these tentative test flights she had remained firmly anchored to her elder sister. Haise and Fullerton took charge of the Orbiter for the third and final Mated Active Test, which reached its climax when the SCA's crew flew the entire pre-release trajectory without a hitch. The next time it would be for real . . . .

FREE FLIGHTS: *Orbiter manned, vehicles separate.*

This third and final phase of the ALT programme was divided into two stages of flight tests. The first series was intended to comprise five flights with the streamlined tail cone in position. In the second series the ferry pod would be removed to simulate an actual approach and landing as accurately as possible, although instead of approaching the dry lake bed runway at the very steep 24-degree angle that would be the 'norm' for actual Shuttle flights, *Enterprise* would approach at a more managable 12-degree angle.

When the Orbiter finally eased clear of the SCA for the first time — at 08:48 local time on 12 August 1977 — it was a magical moment. The men who got the best view of this historical event were undoubtedly the two-man crews of the T-38 chase planes, whose job it was to study every minute detail of the release operation so that it could be discussed at the thorough debriefing session that would follow the initial euphoria. With two long white vortices streaming gracefully from her wingtips, *Enterprise* pulled away from the vast silver mass of the Boeing, descended rapidly through the downwind phase of her landing approach, made a gentle

The momentous 'rollout' of the first Shuttle Orbiter, *Enterprise,* in September 1976. An audience of 2,000 NASA, congressional and industry guests heard the vehicle hailed as the start of 'a new era of transportation.'

180-degree turn onto 'finals' and touched down with seemingly consummate ease. Astronauts Haise and Fullerton looked at one another and smiled — the cynics had been silenced. This "turkey" *could* fly!

Only two more free flights were conducted with the big tail cone in situ. The ALT planners decided to cancel the final two flights of this series because all the test objectives had been met by the first three, and NASA didn't have money to burn on mere demonstration flights. Now it was time to remove the ferry pod and see how the combination would react to the heavy buffeting caused by those huge rocket engine dummies protruding out into the airflow....

On 12 October 1977, slowed down by all that extra drag, the ungainly duo managed to struggle to an altitude of 7,680 metres before pushing gently over into a shallow dive in preparation for the spacecraft's eagerly-awaited release. Just 2 minutes and 34 seconds later, *Enterprise* was back on *terra firma* — in one piece. Two weeks later, the pick-a-back flight test procedure was repeated, again *sans* tail cone and again with total success. The thirteen-flight ALT programme was over. NASA had proved that their shiny new spacecraft really could glide back to Earth after re-entry for reuse at a later date.

With the ALT flights safely behind them, NASA could now concentrate on testing the Shuttle vehicle as a whole in various configurations, including that of a total entity — the Orbiter, solid rocket boosters and external tank functioning as a single unit. This phase of the programme involved a series of rigorous shake-tests and dynamic loading tests designed to simulate a powered ascent from the launch pad. But all this would be achieved *indoors* — quite a feat in itself when one considers the astronomical loadings involved — by installing the Shuttle in a gigantic building at the Marshall Space Flight Center in Huntsville, Alabama, originally constructed for dynamic testing of the Saturn V moon rocket. Before *Enterprise* could be flown atop the SCA for such a long journey, however, NASA felt that a brief series of long-distance test flights should be carried out to ensure that there was no reason why the combination should not remain airborne for extended periods.

In mid-November 1977, a total of four long-duration flights lasting between 3¼ and 4¼ hours were flown in the vicinity of Dryden Flight Research Center, along a route which allowed access to nearby runway facilities should an emergency arise *en route.* But, true to form, all was well.

A bizarre spectacle greeted onlookers near Rockwell's Palmdale plant in 1979, when *Enterprise* was towed 35 miles overland following a series of tests.

Enterprise's journey to Marshall SFC on top of the SCA included a three-day stopover at Ellington AFB, Texas, just a few miles from Johnson Space Center. The idea of this was to allow personnel employed at JSC to see the spacecraft 'in the flesh'; many of them were seeing the machine for the first time.

On arrival at Marshall SFC, the Orbiter was demated from the SCA and installed in a very special test rig known as the MVGVT (Mated Vertical Ground Vibration Test) stand. The first phase of these tests involved attaching *Enterprise* to an external tank, the aim being at best to confirm — at worst to correct — mathematical predictions relating to the vehicle's performance after SRB separation two minutes into an actual Shuttle mission. To achieve this, the Orbiter/tank assembly was suspended in a near vertical position from an air suspension system which included a large truss, air bags and cables. The external tank's liquid oxygen tank was filled with progressively smaller quantities of deionised water to simulate the gradual consumption of propellant that would occur during a real ascent as the Orbiter's main engines slaked their thirst.

When the multitude of instrumental sensors indicated that all systems were performing basically as predicted, the technicians at Marshall knew the second major phase of the MVGVT programme could begin. The Orbiter/external tank combination was removed and two SRBs filled with inert propellant were installed and shock-tested to simulate conditions during the first moments of launch and the long series of tests were repeated.

Finally, a pair of empty SRBs were attached to the Orbiter/external tank combo to verify predictions relating to that phase of a Shuttle mission when the boosters are about to separate. Completed by late 1978, these and other structural tests performed on all the Shuttle's major elements provided the vital endorsement needed before the Space Shuttle could progress towards its maiden space flight . . .

Orbiters are assembled by Rockwell International's Space Division at their factories in Downey and Palmdale, California. The spacecraft is roughly the size of a DC-9 airliner, but there's so much more to it than even the most sophisticated aeroplane. Each one has five on-board computers, bristles with up to twenty-three antennae — and has forty-nine rocket engines!

Although it had been designed to perform a unique role, its structure is in fact fairly conventional. Most of the airframe is fabricated from aluminium alloy panels, frames and bulkheads. The stubby 'double-delta' wings are built by Grumman at their Bethpage, New York facility and feature a corrugated main spar and truss-type internal ribs and stringers. There are wells for the main undercarriage and glove fairings where the wings meet the fuselage. The vertical tail assembly is the work of Fairchild at Farmingdale, New York and is of a similar construction, but with two spars. It is bolted directly to the aft fuselage structure and features a rudder/speed-brake assembly comprising upper and lower sections formed from an aluminium honeycomb material.

***Enterprise* makes a slight turn and bank manoeuver during the second free flight of the Shuttle Approach and Landing Test conducted on 13 September 1977.**

Protruding out directly beneath this tailfin are the three huge conical nozzles of the Space Shuttle's main engines, or SSMEs. In the early development days, these were one of the big headaches of the Space Shuttle programme, but the revolutionary new powerplants have since proved themselves to be both doggedly reliable and ruthlessly efficient.

It is the cluster of three SSMEs that join forces with the solid rocket boosters to propel the Shuttle vehicle up off the launch pad and onward into near-orbital flight, before a system of smaller orbital manouvering engines takes over. The SSMEs ignite at 120-millisecond intervals just before launch, and reach their rated power level of 375,000 pounds static thrust in only 4.5 seconds. During ascent, the engines are gimballed to provide the pitch, yaw and roll control necessary to obtain the correct trajectory. At the point where the Shuttle is undergoing the maximum stresses and strains as the speed of sound is approached — a period known as 'Max Q' — the SSMEs and the solid rocket boosters throttle back to stop these external forces from crushing the vehicle. In the case of the SSMEs, this throttling back process involves a cutback of 65 percent of rated power output for a period of twenty seconds. After the solid rocket boosters have parted company with the craft, the Orbiter's main engines continue to guzzle liquid oxygen and liquid nitrogen from the massive external tank at a mixture ratio of six to one (by weight) until the fuel supply is exhausted. Soon after, approximately eight and a half minutes after launch, MECO (main engine cut-off) occurs and the SSMEs' job is over for the duration of the mission.

This logo (in red white and blue) was chosen as the official insignia of the ALT programme.

Such a mission profile, though simple to relate here, actually presents a formidable challenge. The SSMEs operate at greater temperature extremes than any mechanical system in common use today. The liquid hydrogen fuel is, at −423 degrees F, the second coldest liquid on Earth. Yet when the liquid hydrogen and the liquid oxygen are combusted, the temperature in the main combustion chamber reaches 6,000 degrees F — higher than the boiling point of iron!

Creating effective seals within a structure capable of sustaining such extremes of temperature was, to say the very least, a major design problem to overcome. And to make such a structure capable of up to twenty uses between overhauls, and with a total lifespan of fifty-five missions, seemed to many observers nothing short of impossible. Nevertheless, the problems encountered *were* overcome, and the results have been mighty impressive. One of the technicians I spoke to at KSC pointed with obvious pride at the three conical exhaust nozzles nestling in the aft end of the Orbiter and put the performance of the SSMEs in perspective like this; "If your automobile engine could achieve the efficency of these babies it would only need to be the size of your fist — and that's a fact. After *Columbia* came back from her first flight I took a close look at the engines to see how they fared. Talk about efficiency. You couldn't even tell they'd been fired! They were as clean as a whistle.

They're clean-burning, and there's virtually no friction wear. It looks like they're going to have very good longevity''.

Clean-burning they certainly are. Any orange flame you may have seen omitting from the engines on television pictures transmitted just prior to launch is due to the burning of foreign matter in the surrounding air and in the vicinity of the flame bucket beneath the launch pad. The combustion product of liquid oxygen and liquid hydrogen is pure steam only. Part of the secret behind this capacity for clean burning lies in the unique combustion cycle utilised by the SSMEs. The liquid hydrogen (which serves as the fuel) and liquid oxygen (the oxidiser) flow from the external tank to be partially combusted in dual preburners, which convert them into hot, high-pressure gases to drive immensely powerful turbopumps. The propellants are ignited by devices not unlike automobile spark plugs, with combustion self-sustaining after ignition has taken place. A turbopump blasts these gases through a main injector unit into the main combustion chamber where combustion is completed. It is the partial combustion of the propellants in the dual preburners which ensures maximum performance by eliminating parasitic losses. The high level of efficiency achieved by the turbopumps also contribute to the SSMEs' impressive performance. The hydrogen fuel turbopump, for example, weighs approximately the same as the V8 engine of a modern automobile, but — with 63,286 horsepower on tap — it develops 310 times the output!

Modified airline maintenance procedures are being used to service SSMEs without removing them from the Orbiter between flights. Most engine components can be replaced by line replacement units if necessary in the Orbiter Processing Facility at KSC, without the need for extensive engine recalibration or 'hot-fire' testing. These radical departures from previous spaceflight maintenance procedures are intended to permit the fast turnaround regime that will be essential for high-frequency Space Shuttle operations later this decade. Prime SSME contractor is the Rocketdyne Division of Rockwell International. Rocketdyne are currently working on a number of measures designed to increase the performance of the SSMEs still further, thus permitting even greater payloads to be carried into orbit.

Directly beneath the Space Shuttle's main engines is an aft body flap, which controls the spacecraft's pitch in atmospheric flight and shields the SSMEs from the heat of air friction during re-entry. With those mighty powerplants working overtime during lift-off, massive loads are transferred into the Rockwell-built aft end of the Orbiter, so this area is reinforced with an internal truss structure built up from titanium strengthened with boron epoxy. This structure passes some of the thrust loads on to the external tank, but much of the stress is passed on to the Orbiter's mid-fuselage section — built by General Dynamics at San Diego — which is the vehicle's primary load-bearing component. The mid-fuselage section also houses the payload bay with its two massive doors and the nimble robotic arm — the 'remote manipulator system' — which is Canada's contribution to the Space Shuttle programme. A novel feature of the Orbiter's rounded snout — also built by Rockwell at Downey — is the 'floating' crew capsule, which is set between upper and lower forward fuselage sections as an independent unit. This arrangement is designed to help damp out launch and re-entry/landing shocks.

***Enterprise* pictured seconds after its first departure from the 747 carrier aircraft, 12 August 1977. The Orbiter plunges earthward faster than a falling skydiver.**

The two bulbous pods situated atop the rear fuselage on either side of the vertical fin are fairings for the Orbital Manoeuvering System, or OMS, engines used for orbital insertion soon after launch and for executing all major orbital manoeuvers, including those necessary to achieve re-entry. These are surrounded by a series of smaller reaction control thrusters designed to alter the spacecraft's altitude whilst in orbit.

To find out a little more about the Orbiter, I visited Dr. Aaron Cohen — then Orbiter Project Manager, now Director of Research and Engineering — in his large airy office at Johnson Space Center, just outside Houston, Texas. As Orbiter Project Manager, Cohen was responsible for the design, development and testing of the vehicle and the dollars associated with those responsibilities. His office was a cool refuge from the blistering sunshine, but it probably isn't always such a pleasant place to be; I learned later that in the short time I'd spent with Cohen, his secretary held off fifteen long-distance phone calls . . .

People at Johnson Space Center speak in awe about Cohen's work output. In hushed tones, one employee told me how he often sees Cohen walking into his office in the morning with a briefcase in each hand! Cohen himself sees this as the logical way to get things done. "I usually get into the office around 7.30 — that's 8.30 KSC time, so I'm here in time to talk to the Cape and to my friends in Washington *(NASA headquarters)*. Usually I go home between 7.00 and 7.30 in the evening — that's 5.00 or 5.30 on the west coast, so I'm able to talk to them until they're ready to go home. On top of that I come in on

The 747 SCA with *Enterprise* aboard. Note additional stabiliser fins on the 747's tailplane to aid the main tailfin, which is partially 'masked' by the spacecraft.

This photograph — taken from a T-38 chase aircraft — captures the precise moment of separation of the Orbiter from the 747 SCA.

Saturdays for at least a half-day. It depends to some degree on what's happening with the vehicle. There's always something comes up. For example, at the moment we're trying to get the Orbiter out of White Sands *(a back-up landing site in New Mexico),* so I was still talking on the phone to the people there at about ten o'clock on Saturday evening, and I was on the phone most of the day yesterday — Sunday — discussing another problem that had arisen. So it's fair to say that most of my waking hours are tied up with the Shuttle program in one way or another''.

But such dedication is not enough in itself. To enable him to maintain command of such a vast project, Cohen has a huge number of people working under him. There's an engineering organisation, a test organisation and numerous operations centres across the country. Cohen; "I have broken avionics out separately — there's a special avionics organisation working for me which handles the Orbiter's avionics system and also the softwear associated with that. I also have responsibility for the remote manipulator system, although that's actually built by the Canadians. My office manages the Canadian effort, in assocation with the astronauts who have to operate it up in space. And finally I have a business management organisation''.

How does he keep tabs on such a mammoth operation? "To work with the Orbiter prime contractor my people basically have counterparts out at Rockwell dedicated to engineering, manufacturing, avionics and so forth, and they have daily discussions with their opposite numbers. They bring me the issues and those issues are usually resolved in what we call a technical status meeting. Meanwhile I talk over the telephone to my *direct* counterpart — who's the Orbiter Project Manager at Rockwell — about three or four times a day. Most of our communications are by telephone, but once a month — whether there is call for it or not — we have a sit-down session at Downey, where we go through the total program. We call it an Orbiter Management Review and we go through the total technical schedule and business management part of the program. We also have to have a very close collaboration with the people at Kennedy Space Center in Florida. That's basically how it's done.

I do also have sub-system managers here at the Johnson Space Center that are responsible for such items as the auxiliary power units or that take responsibility for a fuel cell, or for a computer. I have a briefing from them once every week at a noon status meeting, when they stand up and give me a status report on how their own particular systems are doing, both technically and schedule-wise, so I'm able to keep abreast of developments that way also''.

I asked Cohen how they first approached the task of designing and building the Orbiter. "The Orbiter is really three things. It has to operate as a booster on launch, then it has to function as a spacecraft in orbit, then it has to perform as an airplane on landing. So it has to be equipped with all the systems associated with those three modes of operation. When we started, we knew that we had an avionics development program to go through, and we knew that we had a thermal protection system development program to go through. We also had weight constraints of course. We had to build an Orbiter vehicle that was between 150,000 pounds and 160,000 pounds dry weight, when all our data said that really we'd need 170,000 pounds dry weight. So when we started the program the first thing we had to do was to cut back from 170,000 pounds to a goal of 150,000 pounds. As things are working out, it looks like we're going to achieve an actual figure of about 155,000-160,000 pounds by the time we've built the first three operational Orbiters.

"The two really big hurdles which presented themselves in the early days were the avionics and the thermal protection tiles. But coupled with that you always have budgetary issues to contend with — trying to meet our technical requirements, trying to meet our schedules, but keeping within the budgetary restraints that had been set for us.

"As far as the other systems are concerned on the Orbiter, there was a lot to contend with. There's the main propulsion system,and the cooling system, and the hydraulic system, and the mechanical systems — which include the payload bay doors and all the mechanisms associated with them — then there's the communications systems, the fuel cells, the auxiliary power units. All of those systems are very complicated, and of course we did have a few problems getting them to work properly, but nothing that I would say was out of the ordinary in getting an ultra-high-technology program moving. You have to realize that the Orbiter is probably the most complicated system that Man has ever put together. Just to give you an example; I know you're familiar with Apollo. *(Cohen was the Program Manager responsible for Apollo's Command and Service Module, which was a major component of the craft that made the trips to the Moon.)* In terms of complexity, the whole Service Module was only the equivalent of the Orbital Manouvering System pods at the rear of the Orbiter. So there you have a situation where one little element of the Orbiter is as complex as an entire module on Apollo."

When the Orbiter re-enters the Earth's atmosphere after an orbital mission, friction between its external surfaces and the surrounding air produce temperatures well above the melting point of steel along some portions of the spacecraft's skin. To shield the Orbiter and its occupants from such searing heat NASA had to develop a thermal protection system that would provide adequate insulation for mission after mission. Earlier manned spacecraft were protected during re-entry by shedding glowing bits of a heavy, resinous heat shield through a process known as ablation. This material vaporised very slowly, carrying away the heat. But this type of heat shield, like the spacecraft it insulated, was designed to be used only once. For the Space Shuttle, once was not enough. With a design life of five hundred missions, the Orbiter would need a thermal protection system capable of flying into space and back at least one hundred times before being replaced. Such a demanding requirement meant that entirely new materials had to be invented. After considering a number of candidates and carrying out innumerable stringent tests, NASA selected four basic materials, each one designed to insulate the Orbiter's aluminium skin against the range of temperatures expected on that particular area of its surface during re-entry.

The Orbiter's nose cap and the leading edges of its wings are the hottest areas during re-entry. When maximum heating occurs, about twenty minutes before touchdown, temperatures on these surfaces reach as high as 1,649 degrees C (3,000 degrees F). A material known as reinforced carbon-carbon, an all-carbon composite, light grey in colour, insulates these critical zones from these astronomical temperatures. Fabrication of the reinforced carbon-carbon, or RCC, structures begins with graphite cloth, which is impregnated with a special resin. Layers of the cloth are then laminated and cured, then treated to convert the resin into carbon. After further processing the material is treated with a mixture of alumina, silicon and silicon carbide to give it its distinctive greyish oxidisation-resistant coating. The Orbiter's nose cap is fabricated as a one-piece component in this way, while each of the wings has twenty-two panels of RCC running along the leading edges.

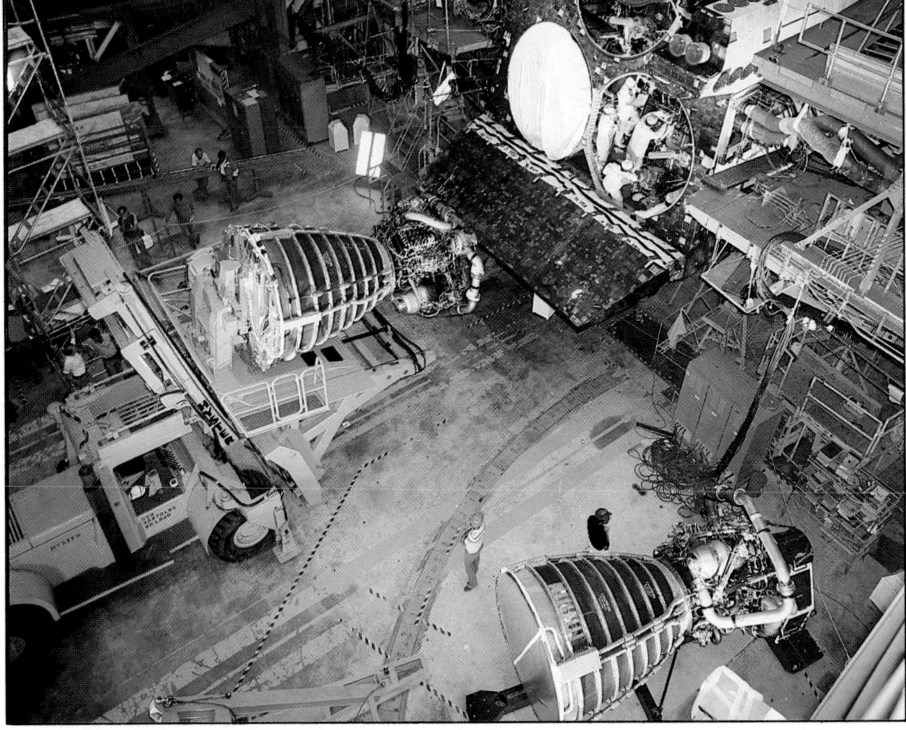

The Shuttle's main engines. Their tiny turbopumps develop 63,286 horsepower apiece!

About seventy-five percent of the Orbiter's external surface is shielded from heat by a system of 30,671 individual tiles — some black, some white — which are formed from a silica fibre compound. These were a source of much controversy over the first few years of the Shuttle construction programme, but their design has since been vindicated by consistent success. Surface heat dissipates so quickly that an uncoated tile removed from an oven only moments before — its interior glowing red with enough intensity to cast light in a dark room — *can be held by its edges with an ungloved hand.* Specially-coated black tiles known as High Temperature Reusable Surface Insulation, or HRSI, cover the bottom of the Orbiter, parts of the vertical stabiliser, and areas around the cockpit windows. With their ability to withstand temperatures as high as 1,260 degrees C (2,300 degrees F), these black tiles are applied where re-entry temperatures are expected to range from 649-1,240 degrees C (1,200-2,300 degrees F).

The white tiles, on the other hand, are designed to insulate the Orbiter from temperatures up to about 649 degrees C (1,200 degrees F) and are applied to the sides of the fuselage, most of the vertical stabiliser, the Orbital Manouvering System engine pods and the upper wing surfaces. They are known as Low Temperature Reusable Surface Insulation, or LRSI.

The manufacture of both types of tiles begins with fibres of pure white silica, which are refined from common sand. The fibres are mixed with deionised water and other chemicals and poured into a plastic mould, where excess liquid is squeezed out of the mixture. After the damp blocks are dried in a microwave oven, they are sintered in another oven at 1,288 degrees C (2,350 F). This fuses the fibres without actually melting them. Rough cutting and precision sizing of the tiles is done with saws, final shaping

A study in unity. Space Shuttle Orbiter and SCA Boeing 747 in their mated configuration.

The Shuttle Orbiter's thermal protection tiles were flight-tested on the wings of an F-15 research aircraft. They were subjected to one and a half times the dynamic pressure that the Shuttle attains during launch. Other tests were conducted on an F-104 Starfighter.

of the surface being accomplished with rotary profile grinders and diamond-tipped cutters. The tiles are then spraycoated, glazed and finally waterproofed. Both types of tile are the same, except for their external coating. This is primarily amorphous borasilicate glass. It is the addition of various chemicals which give the tiles their different colours and heat-rejection capabilities, depending on the re-entry temperatures they will have to withstand.

But for all its resilience to heat, the tile material is still quite brittle, so it has to be isolated from the stresses on the Orbiter's aluminium structure during flight. Launch blasts, aerodynamic pressures, steering forces, thermal stresses, vibration and acceleration cause the vehicle's body to bend and its skin to shift slightly during the ascent into orbit. Once in the cold of space the vehicle shrinks slightly, only to expand and bend again in the searing heat of re-entry. To prevent damage to the brittle tile material, strips of felt padding known as strain isolator pads are used between the tiles and the Orbiter's skin. They are bonded to both the tile and the Orbiter with a red room-temperature vulcanising cement with a rubber base.

Engineers at NASA and Rockwell Space Systems Group also developed a process of strengthening tiles by densifying the back surface. A 'wet cement' slurry composed of ammonia-stabilised binder mixed with silica particles is brushed onto the back surface of the tile in several coatings, penetrating about one-tenth of an inch. This treatment increases the strength of the tile — at the surface where it bonds to the strain isolator pad — by a factor of two. Greater strength is also achieved in some black tile material by increasing the density of the silica material. While most of the white and black tile material weighs four kilograms (9 pounds) per cubic foot, in some critical areas where greater strength is needed, tile weighing ten kilograms (22 pounds) per cubic foot is used. The processing and inspection of each tile is documented and individual tiles are traceable back to material lots.

A final type of thermal protection material used on the Orbiter is a flexible Nomex insulation coated with a white silicon material. Known as Felt Reusable Surface Insulation, or FRSI, it protects external surfaces from temperatures between 177 and 399 degrees C (350 and 750 degrees F). The blanket-like FRSI is used on about 25 percent of the vehicle, covering the payload bay doors, the sides of the upper fuselage, and those upper wing surfaces situated some distance from the wing leading and trailing edges.

Development of the Orbiter's thermal protection tiles proved to be one of the greatest technical challenges of the whole Space Shuttle programme. The initial controversy which first surrounded them was perhaps inevitable. Here was a revolutionary new system designed to be tested for the very first time with the lives of two men at stake. Space Shuttle was the first American spacecraft to be flown on its maiden flight with a crew aboard. All other systems have flown unmanned first, in accord with NASA's traditional "better safe than sorry" approach. After their sudden

Cutting over 8,000 miles of grooving in the Orbiter runway at Kennedy Space Center to prevent the spacecraft from hydroplaning after a heavy Floridian downpour.

failure in laboratory tests, the world's press seized on the tiles issue and blew it up just as far as it would go.

Aaron Cohen knew the problems better than most. "The thermal protection system was very much a state-of-the-art process, as compared with the usual expedient of using the type of once-only ablative system that we had employed on previous manned spacecraft. We had to go with a one hunded-time reusable system that would weigh only the same as balsa wood. The old ablative material weighed about 100 pounds per cubic foot, as compared with the 9 pounds per cubic foot of the Space Shuttle's tiles. Quite a difference!

"We had thought at the time that our biggest development problem was going to be with the leading edges of the Orbiter's wings — the reinforced carbon material — but as it turned out, that worked very nicely. Our big problem was with the rest of the thermal protection system; not from the point of view of the system's inherent characteristics or its durability, but in terms of our ability to attach it to the vehicle. There was a very subtle issue there that said that the tiles themselves should be strong enough to withstand approximately 13 pounds per square inch in a tensile-type pull. But when we bonded the tiles to the strain isolator pads, which isolate the tiles from the stresses and strains going through the Orbiter's aluminium skin, we found that this reduced the strength of the tiles by approximately half. That was quite a surprise to us.

"Hindsight says we should have been able to anticipate that situation, but we didn't find out until much later, when we had already bonded quite a few tiles to *Columbia.* We had run some earlier development tests where we took some typical tiles and bonded them to a typical structure and put it in wind tunnels and shock-tested it — vibrated it — and we didn't have that type of problem. But when we started doing controlled laboratory tests we found that we had a very wide distribution curve of loading effects. It was completely different from the situation you get when you are working in the lab with metal, when you get a very narrow distribution band over a number of samples, with very little discrepancies. We found that with the ceramic material bonded to the strain isolator pad you get a very wide distribution; some tiles had a very low level of strength, and some had a very high level of strength. It wasn't a simple thing to determine the loads that each tile receives. In different areas of the Orbiter you get different aerodynamic shock levels and you have different thermal gradients through the structure. There are different vibro-acoustic levels, different aerodynamic loads. It was really pushing the state-of-the-art to understand how to determine what kinds of loads a given tile actually saw.

"That all hit us right at the same time as our other problem, so there we were with the tiles half the strength we originally thought and the actual loads involved very hard to calculate. Necessity being the Mother of Invention, we did eventually come up with a technique which solved the problem; the so-called densified tile. Bearing in mind that the ceramic tile material is porous, essentially what we do is to impregnate that part of each tile which is bonded directly onto the strain isolator pad with a specially-fomulated substance which hardens when it dries. This way you get the inherent load-bearing capability of the tile itself, rather than of its bond.

"At about the same time we also devised a method by which we could measure the strengh of each tile, despite the fact that the tile material itself has non-uniform properties. Basically, what we did was to pass sonic waves through the tiles before they were bonded onto the vehicle, thus enabling us to screen out the low-strength tiles from the high-strength tiles. That way we could always pick a tile at the high-strength end of the spectrum. We calibrated statistical data that told us what the strength of each tile was by the characteristics it exhibited as the sound waves passed through it. We knew which ones could withstand six or seven pounds per square inch, which ones were thirteen pounds per square inch, which were twenty pounds and so on, and allocated them a place on the Orbiter's exterior according to the demands we calculated would be made on them. We also developed a technique to check the strength of the bonds by using a series of microphone pick-ups attached to the tiles to help us to understand the noises that were coming from the bonds as the sound waves passed through them.

"In summary, the tile issue was a very complicated one. It took a lot of energy, time and ingenuity from our contractor, Rockwell, from Johnson Space Center engineers, and other NASA centers throughout the country — such as Ames and Langley Research Centers. As things have turned out the tiles have proved themselves to be a very efficient means of preventing the heat from reaching the vehicle on re-entry. It's also a very cost-effective and low-weight system. And although we've found that they are susceptible to local damage, they're very forgiving in this respect and they're readily repaired between flights."

I asked if it was just good fortune that, so far, tiles had only been lost from non-critical areas on the Orbiter during missions. Cohen; "I think you have to realize that we carried out a very detailed assessment of the vehicle, and we know where the critical areas are. Most areas on the bottom of the Orbiter are critical, and these are covered mainly by the densified tiles. Densified tiles have twice the strength of the undensified tiles. Now there are *some* undensified tiles on the bottom of the vehicle, but these are in less critical areas than the densified tiles, and all of them have been proof-tested. So it's not by good fortune that we've only lost non-critical tiles, it's by knowing what we had to do and making sure it was done. We have never lost a densified tile. Actually, we're going to densify *all* of the tiles, step by step as time goes on; roughly between four hundred and eight hundred tiles between each flight. The new Orbiter, *Challenger,* already has all of its tiles densified. There's an overall weight penalty of about 1,800 to 2,000 pounds, but we think it's worth it."

Technically speaking, the Orbiter may be described as aerodynamically unstable, so it needs a sophisticated avionics system to enable the crew to fly it back to earth after re-entry. A digital 'fly-by-wire' system is employed, whereby astronaut commands are converted into electrical signals and transmitted to a series of computers carried aboard the Orbiter for processing. In a matter of milliseconds, the computers relay these commands to the spacecraft's reaction control jets, the rudder and other control surfaces and devices so that they function in a correct manner to fulfil the commands. A 'fly-by-wire' system saves weight and complexity by doing away with the need for crude push-pull control rods and actuation cables running the length and breadth of the vehicle to the various mechanical systems. Similar set-ups are to be found on many new-generation military and civil aircraft. Devices known as multiplexer-demultiplexers, or MDMs, are

Technicians in the Orbiter Processing Facility maintain a one-tenth scale drawing detailing the maintenance status and code number of every thermal protection tile.

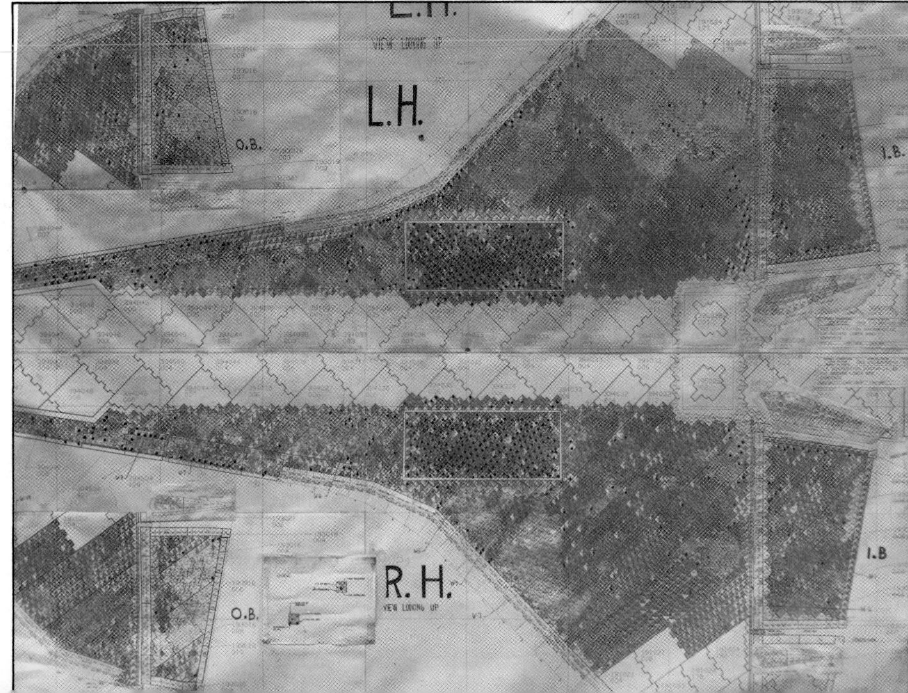

The herringbone pattern of black thermal protection tiles which cover the undersurfaces of the Shuttle Orbiter are shown to advantage in this shot. *Columbia* is being hoisted up to be mated with its external tank and solid rocket boosters.

responsible for transmitting and receiving the commands and their associated calculations. Composed of modular electronic circuit boards which encode and transmit data to and fro, the MDMs permit many different electrical signals to be sent over the same wire, resulting in a substantial weight and cost saving by eliminating the hundreds of miles of wiring that would otherwise be required in an electronic network as complex as that of the Space Shuttle's. Although eight additional MDMs were installed as part of the developmental flight instrumentation for the first few proving flights, operational Shuttles have twenty-three MDMs, including those installed within the solid rocket boosters.

The Orbiter's avionics system is in fact linked to no less than four high-speed digital computers, arranged in a primary redundant (or back-up) set, together with a fifth computer which is essentially independant of the other four. This type of set-up allows the mission to continue even if multiple failures occur. The four computers in the redundant set 'talk' to each other approximately 440 times a second. They get information in and do comparisons and they send information out and do comparisons. One computer is designated as the lead computer, and should this make a mistake then it is voted out of the redundant set by the others.

Controlling the Orbiter's on-board computers are the most sophisticated software programmes ever developed for spaceflight. They not only provide semi-automatic vehicle operations from launch to landing, but also take care of many pre-launch checkout operations. The software programmes are held by NASA's Johnson Space Center, and contain over 500,000 IBM-written instructions. This is *twenty-five* times more than the programmes developed for the Saturn launch vehicle which guided the Apollo astronauts from launch, through orbital insertion and into lunar trajectory.

Cohen; "To lash up all that software and hardware and make it work — and make it work consistently and reliably — was a very large task. Not only was it a complicated process developing the hardware, but developing the hardware and the software to the point where they would function properly together, and then operate in a man-machine relationship, proved very challenging. We had to build very sophisticated labs — such as our Shuttle Avionics Integration Laboratory at Johnson Space Center and the Flight Systems Laboratory at Downey, California — to integrate this system. Well, that was a big hurdle prior to the Approach and Landing Test (ALT) program at the Dryden Flight Research Center, where we launched the Orbiter from the top of the 747 carrier aircraft. But we were really concerned as to whether we were going to make that happen. Up until the last few months we had a very difficult time getting the computers to operate consistently in a redundant set. Eventually they did, and as history shows, the ALT program was extremely successful. In fact, on one flight

Tiles, tiles, tiles.The patchwork quilt of thermal protection materials contrasts sharply with the one-piece reinforced carbon-carbon (RCC) nose cone which must withstand the maximum temperatures seen by the Orbiter during re-entry. The windscreens are a miracle of modern technology, withstanding phenomenal heat variations.

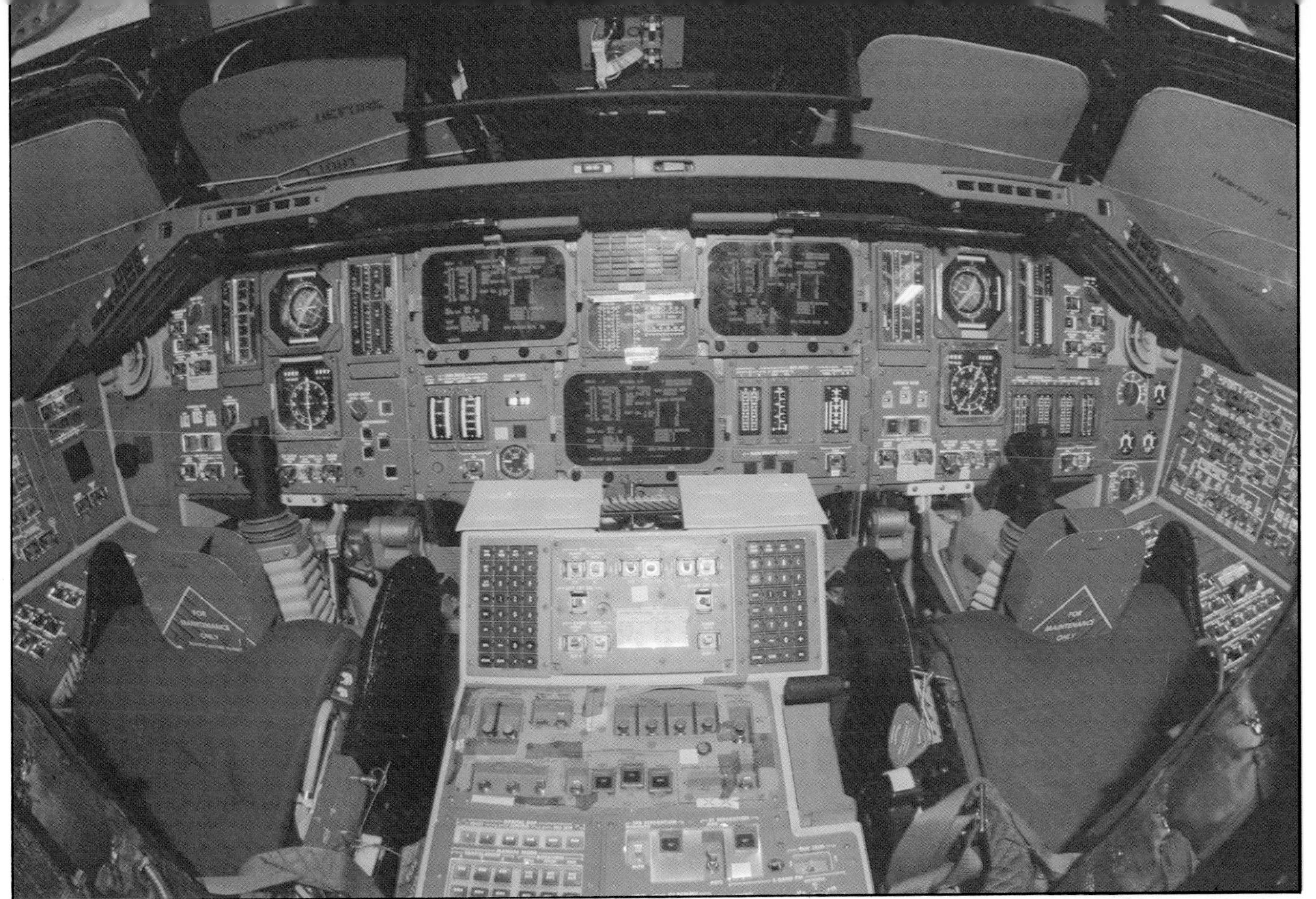

Interior view of the Shuttle Orbiter's flight deck. Detailed instrumentation diagrams are featured on the poster contained in this book.

we had a computer that actually malfunctioned and it was voted out of the redundant set. We didn't do that on purpose, but everything functioned as we had intended it to. That was a tremendous hurdle for us to get over; essentially proving our hard-core data processing/avionics/flight control system".

A small fleet of operational Orbiters is already beginning to take shape. *Enterprise* (code number OV-101) is still used as a ground test vehicle, soon destined for Vandenberg Air Force Base, just northwest of Los Angeles. This will be the USAF's Shuttle operations centre, where military flights — primarily into near-polar orbit — will be launched. Vandenberg will become operational in 1986. *Columbia* (OV-102) was joined at KSC by *Challenger* (OV-099) in July 1982, and *Discovery* came along in October 1983. *Atlantis* (OV-104) will be delivered in December 1984. Three Orbiters will be based at Kennedy and a fourth one at Vandenberg. Funding for the fifth and final Orbiter had just been approved at the time of closing for press.

Challenger and subsequent Orbiters differ slightly from *Columbia,* due to changes in specification which have been made in the light of experience gained on the early test flights. Steady refinement is in any case normal development in the life of a spacecraft. A casual glance over *Challenger's* exterior probably wouldn't reveal any differences, but if you looked a little closer you'd see that there are a number of modifications to the original design. In the case of the thermal protection system, for example, there's now a better tile material, both stronger and lighter than the existing one. It's called Fibrous Reinforced Ceramic Insulation, or FRCI, and it weighs only 12 pounds per cubic foot; its predecessor weighed 22 pounds per cubic foot. NASA and Rockwell have also gone ahead with the installation of a new version of FRSI, which is basically an advanced material capable of withstanding temperatures of around 1,100 degrees, as opposed to the original FRSI's 750-degree capability. They will also replace some of the white (LRSI) tiles on the OMS pods and near the trailing edges of the wings with this new version of the blanket material (it's known as AFRSI) because it not only has a slightly higher capability than the white tiles, but it is much easier to install.

Challenger's interior includes a number of changes in specification, too. Up in the flight deck, the crew now have a head-up display, or HUD. A HUD system projects vital pieces of information up onto a glass panel in front of the windscreen, focussed at infinity so that the crew don't have to look to and from their instrument panels when they want to concentrate on the view up front.

The Orbiter's flight deck features cathode ray tube (CRT) technology. This system 'tidies up' the instrument panel by replacing many individual secondary dials with a single television screen-type display which presents the information the crew require only when there's an immediate need for it. CRTs are a feature of all the best-dressed airliners these days.

The final significant improvement to the Orbiter's specification is the new Ku-band radar system. Rockwell have also made several smaller external modifications, primarily concerned with their efforts to ensure that the hundred-mission tile-life target is successfully reached.

I asked Cohen if budgetary considerations in any way compromised what would have been the ideal Orbiter configuration. His reply was emphatic; "I don't think that the budgetary aspect has compromized the Orbiter's configuration at all. I really don't. Dollars were not allowed to affect either the Orbiter's configuration, or the operational safety standards we set for it. Nor did we allow it to affect the testing of the Orbiter. I think you have to say that the only thing that was compromized — if that's the right word — was the schedule. At one period in the program we had to use the schedule as our variable; in other words we were told that when a problem arose, instead of funding the problem we would pay for it by slipping the schedule. That was the only thing which distinguished the Shuttle development program from the Apollo development program. But we did not sacrifice the safety, quality or capability of the vehicle. No way."

Dollars notwithstanding, was he happy with the way the Orbiter developmental programme was progressing? "I've been on the Shuttle program since August 1972, and I guess I felt it was an honor to be selected for this job. I don't know of anything I could have done that would have been more gratifing to do, though every job has its minor frustrations. But looking back, I can't think of anything I would want to do differently or would want anyone else to do differently. I've got very good staff, I get fantastic support from *(Space Shuttle project chief)* Dr. Kraft, and I guess I'm pretty pleased all round".

There seemed little more to add.

LAUNCH!

The John F. Kennedy Space Center, or KSC as it is more usually referred to, is situated in the area known as Cape Canaveral, on the eastern coastline of Florida about thirty-five miles due east of Orlando. The Center's operational areas are to be found adjacent to Mosquito Lagoon and the comparatively shallow Banana River inland waterway and the extensive marshes all around. Visitors soon find out that the lagoon did not receive its name frivolously, and that a reliable brand of mosquito repellant can be worth its weight in gold!

Here on the fringes of the sub-tropics, the summers are hot and humid, with sudden afternoon downpours which bring a few moments of relief. Even winter is a season of warm days, with snow virtually unheard-of, although occasionally a cold wave sweeps down from the north. The area occupied by NASA is just part of a huge coastal wildlife refuge. Dozens of extremely rare species of birds live here, including the impressive Bald Eagles which can be found nesting not far from the Orbiter's long concrete runway. This magnificent bird — official symbol of the United States — builds its rambling nest with scant regard for the wellbeing of the tree which bears it. Stage by stage they extend their untidy abodes until the sheer weight of material topples the whole tree to the ground! And then there are the alligators — which practically outnumber the people working at KSC. Many of the 'gators have become quite tame, and some have developed a habit of clambering out of the pond in front of KSC's large canteen facility to 'beg' scraps of food from the office workers who eat their lunch outside on the lawn. When the 'gators eventually grow too large for this pastime to be considered entertaining, they are carefully moved to a less populated area off the base.

In this unlikely setting, the pioneering Mercury and Gemini space missions had their beginnings, followed by the even more ambitious Apollo and Skylab missions. It's the only place I've ever been where you can hear someone refer to the Moon and feel like they're talking about the house next door. Although it's situated amidst an endless sprawl of jungle-like greenery, KSC is easy to reach by car. As you cross the stagnant waterways which crisscross the area, the stench of rotting vegetation takes your breath away — but all the time you can feel the old adrenalin racing. The area around Kennedy Space Center really does have that 'certain something'.

To provide a closer insight into KSC's role in the Space Shuttle programme. Rocky Raab of NASA's enthusiastic public affairs team suggested a guided tour of the Center's major facilities. As our car rolled out of the press enclosure I asked Rocky how many people earn their living at KSC. He replied that at the last count there were some 14,000 personnel, about eighty percent of whom work for aerospace contractors, with the balance being employed by the United States government.

A guided tour of Cape Canaveral is a memorable experience. We saw the site where, on 24 July 1950, the very first rocket launch from this area took place — the rocket in question being a captured German V2, modified with the addition of a

General view of Kennedy Space Center, Florida, with the gigantic Vehicle Assembly Building in foreground. The twin-bay OPF can be seen nearby, as well as the long crawlerway that links the VAB with Pad 39A and Pad 39B (extreme left).

U.S. Army 'WAC Corporal' second stage. Since that time there have been many occasions when the eyes of the world have turned to the Cape to watch history in the making. Perhaps the most memorable moment came when Neil Armstrong and Edwin Aldrin set off on the Apollo II mission on 16 July 1969 to become the first human beings ever to set foot on another planet.

Tragedy, too has been part of the Cape Canaveral story. I was shown the spot where, on 27 January 1967, astronauts Ed White II, Gus Grissom and Roger Chaffee perished in their Apollo spacecraft during a fiery training accident. As we wound our way past the historical sites and back towards the KSC of today, one became aware for the first time of a *tradition* being forged here. You could well believe that spaceflight was finally 'coming of age' with the advent of the Space Shuttle era.

The extraordinary Super Guppy transport aircraft literally folds in two to accommodate oversized loads. It is frequently used to carry the Shuttle Orbiter's seventeen-section tail fairing, which is used for ferry flights atop the Boeing 747 carrier aircraft.

We drove back via the operational launch pads on Cape Canaveral Air Force Station (CCAFS), just across the Banana River from KSC, where expendable rockets — tried and tested Deltas and Atlas-Centaurs — regularly send weather and communications satellites into geostationary orbit some 22,250 miles above the Equator. A geostationary, or Clarke, orbit positions a satellite in such a way that it is constantly over one point on the Earth. The last launch of an expendable vehicle from KSC itself, as opposed to CCAFS 'across the water', was the collaborative US/Soviet Apollo-Soyuz mission in July 1975 from Pad 39B.

With a view to tracing the path an incoming Orbiter takes before it can be blasted back up into space, the first stop on our tour of KSC was the three-mile long runway that has been purpose-built to accommodate the spacecraft's high-speed landings. But my impressions of that have been held over until the *Return to Earth* chapter in the interests of continuity.

Rocky and I picked up the Orbiter's trail when we motored along the taxiway that links the long runway with a large concrete apron in front of the so-called 'Mate/DeMate' device. This is the imposing structure used to lift Orbiters off the back of the modified Boeing 747 which ferries them from place to place piggyback-fashion. Because the Orbiter *Columbia* had been transported back to KSC from White Sands just a few days before my arrival, the big Boeing was still there.

The 'Mate/DeMate' device is custom-made. To quote Rocky Raab, "The only thing you can do with it is lift Orbiters on and off Seven Forty-Sevens". The 747/Orbiter combination is towed along a yellow line painted on the apron to ensure that it is correctly aligned with the pale grey superstructure. A bright yellow sling device is then lowered down to within a few feet of the Orbiter, followed by two long platforms which allow access to the sling attachment points on the spacecraft's fuselage sides. Teams of technicians can walk along these platfoms to connect the sling, instead of walking on the delicate thermal protection tiles attached to the upper surfaces of the Orbiter's wings. Once the weight of the vehicle has been taken up by the sling, the attachment points on the 747's support bipods — which have held the spacecraft firmly in place throughout the preceding ferry flight — are unlocked. With this task safely completed, the 747 is slowly wheeled out backwards, and the Orbiter is lowered to within a few feet of the ground. Next, the spacecraft's landing gear is lowered, and the vehicle allowed to stand on its own undercarriage ready for towing to the building known as the Orbiter Processing Facility.

There are only two of these gigantic 'Mate/DeMate' structures in the world; one here at KSC, and the other at Edwards AFB, (where the early proving flights terminated) — later to be moved to Vandenberg AFB. If the Orbiter is forced to land anywhere else — such as the alternative landing sites at White Sands, or at Rota in Spain, or at various other places worldwide — then the 'Mate/DeMate' devices (which are permanent fixtures) are useless. It is in these situations that the 'portable' Stiff-Legged Derrick is employed. This is a somewhat crude, though effective, apparatus; basically a three-legged steel

A barge carrying a Shuttle external tank is manoeuvered by tugboats into the turn basin at Kennedy's Launch Complex 39 for unloading and transport to the nearby Vehicle Assembly Building. Hoisted onto its work stand, the external tank will be mated with the Orbiter and twin solid rocket boosters.

White Sands, New Mexico. The Stiff-Legged Derrick is used to lift Shuttle Orbiter *Columbia* up for the Boeing 747 SCA (see background)...

framework that supports the front of the Orbiter while a powerful crane lifts the rear end. The Stiff-Legged Derrick can be transported at short notice to wherever it is needed by a standard USAF Lockheed C-5 Galaxy transport plane (the world's largest aircraft), 'though even this gargantuan beast must make several trips . . .

After looking over the 'Mate/DeMate' device my attention was drawn to another fascinating sight on the Shuttle landing facility's concrete apron. It was NASA's ungainly-looking 'Pregnant Guppy' — the 'folding' transport aircraft, which literally hinges down the middle to allow it to accommodate oversized loads — which had called in to collect *Columbia's* ferry pod. It's probably fair to say that the Guppy (of which about seven were built) is the ugliest aircraft flying today. It owes its very existence to the American space programme, having been developed in the early 'sixties to fulfil NASA's requirement for an aircraft capable of transporting huge prefabricated sections of Saturn V rockets.

Early Guppies started life as standard Boeing Stratocruiser airliners of the type used by such distinguished operators as Pan American and BOAC in the 'fifties. By removing the upper fuselage and grafting on a bulbous upper extension to provide an interior height of 6.20 metres (20 feet 4 inches) and lengthening the fuselage by over 5 metres (16 feet), then adding a huge hinge assembly at a 'sawn off' point near the front end, the desired effect is achieved. The original 'Pregnant Guppy' took to the air on its maiden flight in September 1962.

To carry even larger rocket sections, a Boeing Stratofreighter was converted in a similar fashion, only this time with the fuselage extended in length by 9.40 metres (nearly 31 feet), and with an extended wingspan and uprated Allison turboprop engines. The result of this metamorphosis was the Super Guppy-201, which first flew in prototype form in August 1965. It is this aeroplane which serves with NASA to this day, the two other 201 models which followed having been based in Europe to carry Airbus A300 airliner subassemblies and large chunks of Concorde between France and Great Britain.

On our journey along the taxiway which links the runway facility and the Orbiter Processing Facility, Rocky pointed out a rather forlorn-looking launch gantry. This huge structure was apparently one of three left over from the Apollo days, two of which were heavily modified for the Shuttle programme. The conventional type of tall, narrow launch tower is not suitable for the rather stocky Shuttle vehicle, so the top two-thirds of what were the Apollo structures were removed and are now mounted permanently on launch pads 39A and 39B to serve as the basis for the more complex Shuttle fixed service structures. The third, unmodified launch gantry now languishing near the VAB is in fact the one used for the very first trip to the surface of the Moon; Apollo II, back in July 1969. In its present form it sports a structure which looks for all the world like a gigantic steel milk stool. Apparently, this modification is an adaptor fitted to facilitate the launch of the shorter Saturn 1B rocket, although the 'milkstool' could be removed at any time to revert to Saturn V launch configuration. This particular gantry was last used for three of the four Skylab missions, and the Apollo-Soyuz Test Project in 1975, the 'milkstool' being retained for all of these flights. Rocky told me there was a plan for the platform of this spare gantry to be kept for future modification to Space Shuttle specification, while the tall tower was to be removed and placed in storage. Initially, the tower was earmarked for use in conjunction with the projected 'Shuttle-derived vehicles', which will use pieces of standard Space Shuttle hardware in an advanced-concept spaceflight system for missions not not requiring a manned capability, (see *A View of the Future* chapter). Current plans, however, are to either scrap the tower, or as some historic groups are petitioning, save it to be reassembled as a monument. Our next destination was the building known as the Orbiter Processing Facility, or OPF. It is to this area that incoming Orbiters are towed once they have been lifted of the SCA 747's back. The processing of an Orbiter vehicle for another flight into space resembles an airline maintenance programme, rather than the time-consuming and highly complex space vehicle checkout and launch operations that were necessary in the Apollo days. For this reason, Orbiters are processed between missions in a building analogous to a sophisticated aircraft hangar.

The OPF consists of two 'high bays' capable of accommodating one Orbiter apiece, connected by a lower section housing electronic and mechanical support systems, workshops and office space. Each high bay is equipped with two 27-tonne (30-ton) bridge cranes and contains a bewildering array of close-fitting maintenance platforms.

Orbiters are slowly towed into the high bays along the obligatory yellow line, their tall tailfins passing through a vertical slot cut into the area above the main doors. Once inside, the maintenance platforms are pulled into position and a small army of elite technicians set to work. Their job is to examine the spacecraft for any faults which may have arisen since the last checkover and test all systems in preparation for the next voyage. Those payloads which are to be processed and installed into the Orbiter's cargo bay in a horizontal position

...to be moved into position beneath it for the journey back to Kennedy Space Center (KSC).

Boeing 747 and Shuttle Orbiter *Columbia* approach the 'Mate/DeMate' device at Kennedy.

Columbia is rotated from a horizontal to a vertical position by means of a special sling before being mated to its external tank in the VAB.

will have these tasks performed in the OPF, while payloads which must remain in a vertical position are mated with the Orbiter out at the launch pad and do not actually enter the OPF. The Spacelab orbiting laboratory will be the most frequent horizontal payload.

On leaving the OPF after being prepared for flight, Orbiters are wheeled into the adjacent Vehicle Assembly Building, or VAB. Here, the spacecraft is mated to an external fuel tank and two solid rocket boosters atop the huge Mobile Launch Platform.

The VAB is the heart of KSC's Launch Complex 39 and has been modified to accommodate the pre-flight assembly of the Space Shuttle's primary elements, having performed a similar task with the massive Saturn V and Saturn 1B launch vehicles back in the Apollo/Skylab/Apollo-Soyuz days. One of the world's largest buildings — and certainly the world's largest *single-storey* building — the VAB stands 525 feet tall and can be seen from many miles away. Its size is such that even when you think you're quite close to it, those gigantic walls — emblazed with the Stars and Stripes and the red, white and blue 1976 Bicentennial symbol — still retain that slightly vague, misty quality of a distant mountain range. In fact, it is no exaggeration to say that the building's colossal box-like exterior is an optical illusion; all previous experiences have taught the mind to accept that really large buildings have a tall, narrow 'skyscraper' shape.

Inevitably, impressive facts and figures concerning this extraordinary building abound. It would be regarded as a gift from Heaven by those who earn their living conducting guided tours — if only tourists were actually allowed in! Apparently, the VAB occupies a ground area of eight acres, and boasts an internal volume of 3,624,000 cubic metres (129,428,000 cubic feet). The structure was designed to withstand winds of up to 200km/hr (125mph), and has a foundation which rests on more than 4,200 steel pilings 40 centimetres (16 inches) in diameter, driven down to bedrock at a depth of 49 metres (160 feet). Inside are four 'high bays', two of which are used to join together the various Shuttle elements prior to launch.

Before being allowed to look around, each visitor is subjected to a thorough questioning and is then frisked by an armed guard. You must wear either full-length trousers or slacks and proper shoes, not sandals or high heels, if you are to be allowed in, and you are told that any matches or cigarette lighters in your possession should be left at the main gate and collected as you leave. After that you are issued with a hard hat, pass through an airport-style metal-detector machine (they detected my tiny portable tape recorder easily) then off you go — an escort by your side at all times. On this particular occasion, my escort was the amiable George H. Diller of NASA's public affairs corps, who remained in contact with his supervisors throughout our tour by means of one of those neat walkie-talkie units.

Once inside, our first stopping point was the area where the various cylindrical segments which make up the Shuttle's solid rocket boosters (SRBs) are prepared for 'stacking' into their launch configuration. Each segment is handled separately until it reaches one of the giant mobile launch platforms housed in another part of the building. George Diller pointed out that, during the Apollo programme, this particular area of the VAB was used for storing some of the 1B and 1C Saturn rocket stages.

The SRBs looked somewhat innocuous just lying there in pieces, but first impressions are often misleading. These slender pencil-shaped tubes provide the highlight of the launch when they lay two long plumes of white-hot flame and thick smoke across the clear blue Florida sky.

A solid rocket motor is inherently simple, requiring no turbopumps and dispensing with the complexity of the propellant feed system necessary in a liquid-fuelled unit. Ignition is achieved by means of a complex sequence of pyrotechnic events at the top of the motor, culminating in the application of a high-temperature flame along the surface of the solid propellant. Despite the complexity of this chain of events, the ignition interval (the time between receipt of an ignition command to the build-up of pressure in the SRB to 563.5 psia) is a mere quarter of a second . . .

The Shuttle's SRBs are the largest and most powerful flight motors ever developed and are the first 'solids' to be built for a manned spacecraft. Each booster is 149 feet long and measures 12 feet in diameter. Together they weigh 2.6 million pounds and produce 5.3 million pounds of thrust at lift-off. The SRBs burn for the first two minutes of each mission when, at an altitude of about 44 kilometres (27 miles), they separate from the spacecraft with the aid of small explosive charges and clusters of tiny rocket motors. Parachutes then lower them gently into the Atlantic Ocean for recovery and subsequent reuse. A twenty-flight lifetime is projected for each SRB.

Prime contractor to NASA for SRB design, construction and development is Morton Thiokol's Wasatch Division, headquartered near Brigham City, Utah, where a series of six spectacular test firings took place between July 1977 and the end of 1980. For these static firing tests, the SRBs were positioned horizontally to fire up a nearby hillside. Morton Thiokol have produced a number of solid propulsion motors for space applications over the years, and currently also participate as a joint-venture partner in the Trident submarine-launched ballistic missile programme.

The SRB production process provides a good example of just how complex the Space Shuttle programme as a whole is, for by looking at the main construction and loading operations involved, some idea of the elaborate interplay of literally hundreds of specialist companies can be gleaned. Work on the SRBs is currently being performed under more than 8,600 Morton Thiokol subcontracts in forty states and the district of Columbia. About half of the contract dollars Thiokol receives from NASA go to subcontractors.

Although at the time of closing for press a new lightweight SRB constructed primarily from composite materials had been given a production go-ahead, for the next few years of operation, the outer case – or motor case as it is called - will continue to consist of eleven cylindrical steel sections, which are later assembled into four main propellant-filled 'casting segments'. These are initially formed and shaped from steel billets by the Ladish Company of Cudahy in Wisconsin. Once formed, the segments are transported to Cal-Doran Metallurgical Services in Los Angeles, where they are heat-treated to strengthen and toughen the steel to withstand the rigors of twenty fiery ascents. The next stage of fabrication involves forming the case tang and clevis joints which hold the case segments together. This operation is carried out by Rohr Industries at Chula Vista, California. Each joint is carefully machined and then drilled to accept 180 joining pins. From Rohr, the case segments are shipped to Morton Thiokol's Wasatch Division to be loaded with propellant and fitted with exit nozzles.

As aforementioned, the SRBs will not remain in their original form for long, because demands for improved performance are part and parcel of virtually any aerospace programme — and particularly space programmes, where weight-saving is so important. With weight-saving in mind, eight of the eleven metal case segments will be replaced with a composite filament material — thus increasing the Shuttle's payload-carrying capacity by about three tons. The new filament-wound cases will be built by the Hercules Company, who are based at Salt Lake City, not far from the Morton Thiokol plant. Hercules have made quite a name for themselves in recent years for their work with composite materials, and achieved a lot of publicity when they became suppliers of carbon fibre monocoque chassis structures to the McLaren Grand Prix racing team.

As well as bestowing the ability to get extra-heavy payloads into orbit from KSC, the new lightweight boosters are needed for high-performance launches from Vandenberg AFB in California. The lightweight motor case will help compensate for reduced lift capabilities when launches from this west coast location cannot take advantage of the extra impetus provided by the Earth's rotation. A series of test-firings at Wasatch will precede the first use of the new composite SRB design in late 1985.

The exit nozzle assemblies situated at the base of each SRB are fourteen feet long and weigh an astonishing eleven tons apiece. During launch they must withstand unimaginable heat levels. Each nozzle is attached to its booster via a flexible bearing incorporating alternate layers of elastomeric rubber and steel to permit the whole assembly to be flexed in flight for guidance control.

The solid propellant used in the SRBs is very strange stuff. Production begins when the fuel (atomised aluminium powder), oxidiser (ammonium perchlorate) and other ingredients are mixed in huge 600-gallon 'bowls' at Wasatch. The resulting material is then poured straight into the casing segments and cured for four days at 135 degrees F, by which time it has acquired the look and feel of a hard rubber

After being hoisted 190 feet over the Vehicle Assembly Building's transfer aisle and moved into High Bay 3, *Columbia* is lowered gently into place alongside its external tank and solid rocket boosters before being mated with them to form a complete Shuttle vehicle atop the huge Mobile Launch Platform.

typewriter eraser. Stringent safety measures are taken at all times, but the solid propellant is in fact both stable and storable in this condition.

It was fascinating to study the aft skirts and other pieces of SRB hardware at close quarters as they lay impassive on their steel pallets. There's a certain tranquility about that area of the VAB — broken only by the occasional echoing tannoy message — which makes one feel rather uneasy. It was difficult to imagine that the aft skirt assembly at my fingertips would soon be supporting an entire Space Shuttle vehicle on the launch platform, and to visualise the drama of lift-off, when a searing column of energy would course through it with unbridled violence.

In this area, teams of skilled technicians incorporate parachutes into the various nose segments — together with the pyrotechnic devices which help to deploy them — and load the explosives which detach the SRBs from the external tank. Further along I saw the small conical nose caps which will sit atop the boosters once they have been assembled, and as we walked on to another section of the building, George pointed out a large door behind which spray-on insulation is applied to the casings to protect them from the tremendous heat generated by friction with the atmosphere as they soar up into the fringes of space.

A hundred yards or so further on we stopped to take a closer look at the big yellow rail cars which bring the SRB segments directly into the VAB after their 2,700-mile, ten-day haul across America from the high plateaus of Utah. They only travel by daylight, and a NASA engineer always rides on board to keep a close eye on everything. The possibility of sabotage must not be ignored.

Another item of interest nearby was to be found in the indoor parking lot for NASA's Payload Transporter Vehicle. This bizarre-looking machine is self-propelled and has *forty-eight* wheels along its length. Its job is to transport the huge environmentally self-contained canister in which the Shuttle's payloads are stored until they are transferred into the Orbiter's cargo bay. Depending upon whether the payload has to be kept in vertical or a horizontal position, the canister can be tilted through ninety degrees to accommodate it.

Apparently, the Payload Transporter Vehicle was built in West Germany and sustained slight damage — easily repairable — during its sea journey to the USA when a storm hit. It's a veritable box of tricks. To aid manouverability all forty-eight wheels are independantly steerable. This permits the transporter to move forwards, backwards, sideways or 'crab' diagonally, or even turn about its own axis like a carousel. There's a driver's cab at either end. The payload canister itself has two huge doors running the length of its upper surface. Out at the launch pad these doors are only opened once an effective seal has been formed with the rotating service structure, which in turn swings across to transfer its precious load into the Orbiter's payload bay. A payload thus transferred is constantly sheltered in a benign environment with such factors as temperature and humidity carefully controlled until the moment it is deployed into the vacuum of space. At no time is it exposed to the outside air.

On the wide expanse of concrete beyond, we once again pick up the trail taken by the Orbiter vehicles as they are prepared for launch. Having been wheeled in from the OPF, the spacecraft must be lifted up to align with its giant external tank and attached firmly to it. The hoisting operation required to lift the Orbiter up into position for mating with the massive external tank is a sight to make your hair stand on end. The spacecraft is towed along a yellow line to ensure precise alignment with a sling device, which is then lowered down from the VAB's ceiling. Once the sling has been firmly located on the Orbiter's flanks, the spacecraft is lifted straight up off the ground and the undercarriage retracted. The vehicle is then rotated into a vertical position and hoisted slowly right up into the roof of the building,

A Shuttle,Pad 39A-bound, aboard the Crawler-Transporter vehicle and Mobile Launch Platform. The white building in the bottom left-hand corner is the Launch Control Center, while the canal and turn basin used by the barges carrying incoming Shuttle external tanks can be seen to the left-hand side. This picture was taken from the roof of the 525 foot high Vehicle Assembly Building.

Columbia is backed out of the Orbiter Processing Facility as part of preparations for the STS-3 mission. The OPF is analogous to a sophisticated aircraft hangar, because Orbiter preflight processing has been greatly simplified to permit faster turnaround times.

then translated across into one of the high bays through an opening set into the VAB's ceiling structure. It's a tight fit; the Orbiter's wingtips clear the opening with literally inches to spare. External tanks also pass through this opening on their way to the high bays for integration with the other Shuttle elements.

The next stage of my guided tour involved a long elevator journey and took us onto one of many maintenance platforms fitted snugly at various levels around the girth of the next external tank destined for launch. These platforms hinge down to allow the external tank to be lifted out of its maintenance cell — there are two external tank cells in the VAB — by one of the huge bridge cranes installed in the VAB's ceiling.

Out on the launch pad the external tank is too big to miss. It stands there like a huge monument to the a aspirations of twenty-first century Man, its graceful lines silhouetted against a glittering Atlantic Ocean backdrop. But here in the VAB, within touching distance of its nobbly skin, I'm too close to appreciate its vast size. It's rather like trying to admire the Statue of Liberty whilst standing at the entrance to her right nostril. The analogy is not so abstract as it might seem, because the tip of that famous lady's fiery torch would stand just two feet above the top of the Shuttle's external tank if — God forbid — both monuments were to be placed side by side. External tanks are manufactured at the NASA Marshall Space Flight Center's Michoud facility in New Orleans by the Michoud Division of Martin Marietta. The company has had a long association with the American space programme, having previously been responsible for such projects as the Viking rocket programme and the revolutionary X-24 manned lifting body.

The Space Shuttle's giant external tank is a sophisticated example of precision engineering. 'Slosh baffles' within it prevent the liquid hydrogen and liquid oxygen propellants from splashing around and throwing the vehicle off course during ascent.

Fully laden — with a total weight of 11 million pounds — the Crawler-Transporter vehicle guzzles fuel at the rate of one gallon every twenty feet!

Completed external tanks are far too large to be transported to KSC on the highways, so they are loaded onto an enclosed ocean-going barge and towed from New Orleans right along the Gulf of Mexico coastline, past the Florida Keys at the extreme south of the peninsula, and then northwards into a short section of canal which juts into KSC. They are moored in a turn basin adjacent to the VAB, a short distance from the KSC press site. In fact, the press site parking lot is actually the offloading ramp for incoming external tanks! Each tank is already mounted on a special trailer, allowing it to be towed into the VAB and hoisted up into a vertical position ready for checkout.

When the solid rocket boosters have expended their propellant and separate at an altitude of 44 kilometres (27 miles), the Orbiter, with its three main engines still burning furiously, carries the external tank 'piggyback' to near-orbital velocity to a point about 113 kilometres (70 miles) above the Earth. There, 8½ minutes into the mission, explosive bolts detonate and the now almost empty tank parts company to fall in a roughly preplanned trajectory towards the Indian Ocean. Friction with denser levels of air as it continues its high-speed fall cause the tank to disintegrate into tiny charred fragments which scatter harmlessly over a wide area.

The external tank is the only major expendable element of the Space Shuttle vehicle. It was considered cheaper to discard it than to strengthen it sufficiently to survive the rigors of re-entry. The external tank is covered with a variety of thermal protection materials. First, a layer of cork is applied to the tank's exterior to prevent 'hot spots' developing that would disturb the smooth flow of liquid propellants into their feed valves. (These hot spots develop when air turbulence created by the Shuttle's unconventional shape generates heat at the very high speeds reached during the latter stages of ascent). Secondly, on top of the cork is an inch-thick layer of spray-on polyiscocyanurate foam insulation to keep the supercool propellants at the correct temperature. This layer also protects the aluminium alloy skin — which is less than ½-inch thick in most places, but over 2-inches thick in others — from aero-dynamically-induced heat and minimises ice formation, which could otherwise damage the Orbiter's delicate tiles by breaking off in chunks during the period of severe vibration which accompanies launch.

There's much more to the external tank than meets the eye. It's exterior looks so inert that it is difficult at first to think of it as anything other than a big steel tube rounded off at both ends. In fact, it's a marvellous example of modern engineering, both inside and out. It must accommodate complex load effects and pressures from the propellants within, as well as those from the two immensely powerful SRBs and the triple-engined Orbiter. As the structural backbone of the Shuttle vehicle in its launch configuration, at lift-off from KSC the external tank must absorb a combined thrust load of 28,580

To avert a potential catastophe, an engineer walks alongside the Crawler-Transporter vehicle at all times, looking out for telltale cracks in vital parts. The Crawler is powered by two 2,750 – horsepower diesel engines.

kilonewtons (6,425,000 pounds). That in itself is no mean feat.

The external tank is actually *two* tanks — one containing 141,000 gallons of liquid oxygen, the other containing 385,000 gallons of liquid hydrogen. These propellants power the three Space Shuttle Main Engines (SSMEs) during the thunderous ascent into the upper reaches of the atmosphere. The liquid oxygen tank only occupies a relatively small space up in the forward section of the external tank's huge structure, with the much larger liquid hydrogen tank behind it (or below it if you are viewing the structure in its vertical position on the launch pad). The hydrogen tank is 2½ times larger than the oxygen tank, but it weighs only one-third as much when filled to capacity. This difference in weight comes about because liquid oxygen is *sixteen* times heavier than liquid hydrogen.

Between the two propellant tanks is a cylindrical structure known as the intertank. In actual fact, this is not a tank at all; it houses vital instrumentation and processing equipment and provides the attachment structure for the forward end of the solid rocket boosters. The utilisation of an intertank structure also makes it possible for the two propellant tanks to have wholly separate domed bulkheads, thus avoiding the design complexity and operational constraints that would be imposed by a common bulkhead configuration. A small door set into the side of the intertank allows ground support personnel to enter this area to carry out checks on the instrumentation contained there, or to clamber through the manhole openings to gain access to the interior of either propellant tank.

There are other clever details beneath the tank's skin. Its interior houses a propellant feed system to duct the propellants to the Orbiter's SSMEs, a pressurisation and vent system to regulate the tank pressure, and an environmental conditioning system to control the temperature and render the atmosphere in the intertank area inert. There is also an electrical system to distribute power and instrumentation signals and provide protection against lightning strikes once the Shuttle is airborne, by means of a cast aluminium rod. (Apollo 12 was struck by lightning during ascent from KSC in November 1969). One notable exception to this self-contained design approach was the early decision to locate most of the fluid control components in the Orbiter rather than the expendable external tank, in order to reduce throwaway costs to a minimum. A sophisticated fuel gauge system in the external tank consisting of a series of strategically-placed sensors constantly checks the quantities of propellant remaining in each tank.

The more you study the external tank's vast structure, the more you realise just how much careful thought had to go into it. For example, a cage-like assembly of circular baffles is built into the liquid oxygen tank to damp out the fluid's sloshing movements during ascent and ensure that this motion does not throw the vehicle out of control. The density of liquid hydrogen, on the other hand, is low enough to render slosh baffles unnecessary. Antivortex baffles are installed in both tanks to prevent gas from entering the SSMEs. Without them the propellants would create a vortex similar to a whirlpool in a bathtub drain. These antivortex baffles, which look not unlike huge fan blades, minimise the rotating action as the propellants flow out of the bottom of the tanks on their way to the Orbiter's engines. It's a very elegant arrangement.

An aerial view of a Shuttle vehicle in position on Pad 39A at KSC. In the foreground is the five-degree incline which the Crawler vehicle negotiates on its way to the launch gantry. Clearly visible is the tall lightning conductor mast atop the Fixed Service Structure and the curved rail for the Rotating Service Structure. This provides access for maintenance and inspection personnel in the days leading up to a launch, but it moves out of harm's way a few hours before the Shuttle's fiery ascent.

After the first two Shuttle flights, external tanks took on a striking new appearance. Their colour was changed from the original stark white to a light brown hue — a transformation which somehow brought an added touch of character to the Space Shuttle vehicle as a whole. But there were sound economic reasons for this sudden transformation. Light brown is the natural colour of the spray-on foam insulation used on the tank's exterior. By leaving off the white paint that had been applied to the tanks used on the first two Shuttle missions, STS-1 and STS-2, a weight saving of 600 pounds was accomplished at a stroke. Since the external tank is carried almost all the way to orbit, this saving translates into almost 600 pounds of additional payload-carrying capacity. And there's a small but useful bonus; a $15,000 saving in white paint! NASA were quick to point out that eliminating the layer of white paint — which was only added for the first two missions for cosmetic reasons anyway — would not affect the foam insulation's fire-retardant or water-resistant properties. On STS-6, a new 'lightweight' external tank entered service, increasing Shuttle payload capacity still further.

My lasting impression of that close encounter with an external tank — since reduced to a few black fragments on the bed of the Indian Ocean — is that the spray-on insulation looked more yellow than light brown, although that may in part have been due to the very intense lighting conditions which prevailed in some areas of the VAB. The random patterning created on the tank's exterior as the foam insulation is sprayed on looks for all the world like the skin of a giant reptile; it's surprisingly rough. Dotted here and there on the skin were tiny black-printed numbers, and blotches of darker orange insulation, presumably where small inspection panels have been covered over with insulation at a later stage.

It was all rather strange; standing there in the upper reaches of the world's largest building alongside an inanimate leviathan that would soon be burning to shreds over an exotic expanse of ocean 10,000 miles away . . .

Next, we took a ride up to the sixteenth floor to get a better view of the 'stacking' operation taking place atop the Mobile Launch Platform, where all the Shuttle elements are integrated. We found ourselves a superb vantage point overlooking the Shuttle vehicle, which stood no more than twenty yards away. We could see a team of technicians clambering into the access doors set into an instrumentation frustrum atop one of the SRBs. That's where all the circuitry is that takes commands from the main computers aboard the Orbiter to gimbal the boosters' aft nozzles for directional control during take-off. From close quarters the Orbiter looks much more complex than I had imagined. The more you looked, the more there was to see. I've spent a lot of time around military aircraft, but the degree of complexity here really was astounding. George noticed my gaping jaw and said, "It may *look* like an airplane — but the similarity stops there".

Even up here, on a level with the conical tips of the solid rocker boosters, the roof of the VAB seemed no closer than it had appeared from the ground floor. You certainly couldn't work here if you suffered from vertigo. From that point of view, the most demanding job of all involves sitting in one of the tiny glass-bottomed control cabins which roll along the overhead beam crane rails 520 feet above floor level!

George broke in again; "See, up there in the upper levels of the VAB they've got those louvres. They're always open in case there's an accident and they need to let the fumes out. It's just to allow those poisonous gases to rise up and escape before they cause further damage. That's a tip-off for people outside the VAB that we've got solids in here. Rather like you flying the Union Jack outside Buckingham Palace when the Queen's in residence!"

Peering down through the safety nets which surround all the maintenance walkways up here I notice that when SRB segments are in a loaded condition — as of course these were — they bear a small black legend stencilled on their sides to say so. George was inspired to punctuate the tour with another interesting tale. "They had an aluminium shed near Pad 39A. I guess it was about fifty yards away from the pad perimeter. It was a prefab building and they weren't sure if it was gonna stand up or not. After the launch you should have seen it. Boy, the Shuttle just *annihilated* it! It was reduced to a few pieces of debris scattered across the ground."

The close proximity of the fearsome

(Above) Astronaut John Young clambers into the 'basket' portion of the slide-wire escape system on Pad 39A during a training session, much to the amusement of Robert Crippen. *(Below)* Astronauts are receiving a briefing on the slide-wire escape system. There are five 1,200-foot long cables and five 'baskets'. At the bottom of each wire is an arresting net, a system of deceleration chains and a concrete bunker.

Safety divers assist a mannequin Shuttle 'crewmember' during a simulated escape from an Orbiter nose section mockup at KSC.

SRBs prompted a brief discussion on their safety. Diller; "It's very difficult to start them accidentally, but in such a confined space once they got going you'd really have a holocaust. When they light them up out on the pad they almost need a rocket within a rocket to get them going. They really need a very high temperature to start them up, so it's unlikely that any of the usual fire hazard-type things would pose any problem". Nevertheless, NASA's long-term plan is that SRBs will not be moved into the VAB until just before the stacking operation begins, thus alleviating the safety problems associated with storing the boosters in such a busy place.

While safety standards are zealously upheld, security in the VAB is also very tight. A series of coloured badges displayed prominently on the clothing of some workers in the building denotes the security rating they have been accorded by NASA. Apparently this is know as the PRP, or Personnel Reliability Profile. Under the PRP security scheme key employees at KSC are carefully screened to ensure that their personal background shows no tendency towards terrorism or acts of irresponsibility that might jeopardise safety. Although it is fair to say that a lunatic let loose near an 'armed' Space Shuttle could cause a disaster of cataclysmic proportions — and that *any* measures taken to avoid this are therefore justifiable - there's also a slightly sinister connotation to the letters PRP. One employee at KSC put it like this; "There *are* certain benefits to gaining a PRP status in terms of getting access to things, but I'd rather stay as I am. PRP is too good an opportunity for Uncle Sam to start snooping into your personal affairs."

"We never had anything like this in the Apollo days", another man told me. "When they get the computerized security system in here it's gonna be a whole different ball game. They'll just man-load an area for the minimum number of personnel required to perform a given function and unless you've been specifically approved to be in there for a certain reason at a certain time you just won't get in".

For me, this preoccupation with security illustrates two conflicting desires on NASA's part. On the one hand there's the need to make space travel routine — to present the impression that this is *everyone's* Space Shuttle. On the other hand there is the very real threat of terrorist activity, not to mention the more mundane limitations imposed by the kind of safety requirements which inevitably accompany inflammable fuels, toxic gases and charges of high explosives — all of which necessitate restriction of access to all but a privileged few. Quite how this dilemma will be resolved is anyone's guess, but at the moment NASA seems to be gravitating towards the 'restricted access' approach. To say that safety standards in the VAB are stringent would be something of an understatement. Notice boards dotted all around the top work levels of the building near the Shuttle stacking area read;-

REMOVE PERSONAL ITEMS FROM THE UPPER POCKETS. WEAR BADGES INSIDE SHIRT OR CARRY IN PANTS POCKET. REMOVE WATCHES AND RINGS OR TAPE SAME. EYEGLASSES MUST BE TETHERED. ALL TOOLS MUST BE TETHERED. ACCESS RESTRICTED WITHIN THREE FEET OF VEHICLE-CONTACT ACCESS CONTROL MONITOR. FOOD AND BEVERAGES PROHIBITED. FLAMMABLE LIQUIDS MUST BE APPROVED. NO HARD HATS ALLOWED.

All this in addition to the screening that takes place before you are even allowed *into* the VAB! The final sentence on those notice boards is somewhat strange, because the wearing of hard hats is compulsory nearer ground floor level. Is

Sharply defined against the night sky, a Shuttle vehicle is illuminated by the world's most powerful system of searchlights. Clearly visible in this shot is the crew ingress/egress 'white room' — alongside the cockpit area — which swings clear prior to launch.

one to assume that NASA consider it more desirable that an unhelmeted technician be struck by a falling object, than for the delicate Orbiter vehicle to be struck by a falling hard hat?

The final part of our VAB tour took us somewhat closer to ground level and involved a walk around the Mobile Launch Platform itself. This particular platform was being prepared for its first Shuttle launch (STS-6), the first five missions having set off from its sister platform. I asked about the modifications that were made to prepare these ex-Apollo launch platforms for Space Shuttle operations. Apparently, their upper decking was reinforced, and extra exhaust ports had to be cut into them. Exhaust ports allow the immensely powerful columns of thrust from the rocket engines to flow down into the 'flame bucket' beneath the concrete launch pad, to be diverted out into the open air some distance from the spacecraft where they can do no harm. The single exhaust port for the main engines of the Saturn V and IB — which were clustered close together — were relocated for the Orbiter's main engines, and two additional holes had to be cut for the searing columns that emit from the SRBs.

Other major modifications to the Mobile Launch Platforms were carried out *after* the first Shuttle launch. These centred around the acoustic suppression system, which sprays 300,000 gallons of water onto the top surface of the platform in a mere fifteen seconds at launch, to help prevent the enormous shock waves from being reflected back up to damage the Orbiter. This water is delivered by conical castings of high-strength structural steel known as 'rainbirds'. There are six of them on each Mobile Launch Platform, and they each stand about twice the height of an average man. George interrupted my thoughts; "You should see one of these things in action during a test. Man, it's like Niagara Falls!" The continuous column of water shoots up the inside of the 'rainbird', strikes a massive deflector plate at the top and cascades out across the launch platform in an almighty deluge. You get some idea of the water pressure involved when you see the size of the nuts and bolts which hold the 'rainbirds' together. The maximum rate of delivery is quoted as 900,000 gallons per minute — resulting in a yard-thick cushion of water over the entire platform. That cushion is absolutely essential, because such powerful levels of low-frequency noise don't do the spacecraft any good, and it isn't particularly beneficial to the payload or the crew either.

(Above) Traditional astronauts' breakfast of steak, eggs, coffee and orange juice for Young and Crippen before STS-1. *(Below)* Suiting-up — in a pair of old armchairs!

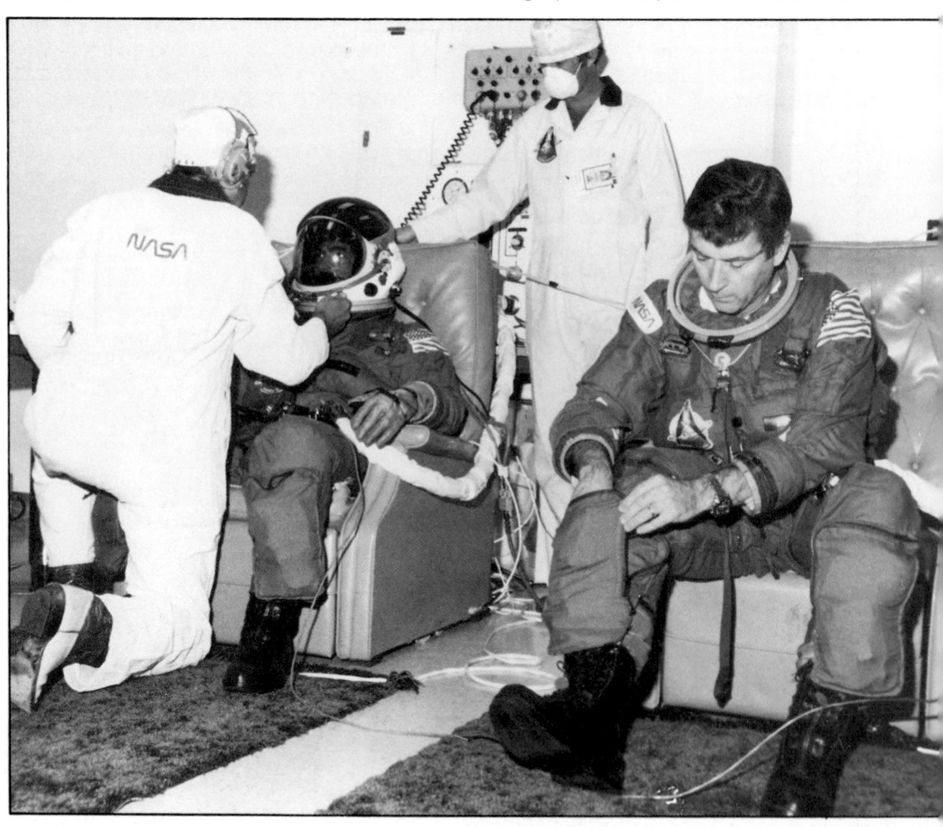

George provided some background to the modifications made to the acoustic suppression system after STS-1. "When you watch the close-up film that was taken of the Orbiter at the moment of take-off on that maiden flight, you can see the control surfaces at the trailing edges of its wings being deflected by the sheer force of overpressure from the impulse created by the solids igniting. So after the first flight one of the main modifications they had to make was to install a water suppression system inside the SRBs' launch platform exhaust ports. The idea of the extra suppression system was to lay a curtain of water across the SRB exhaust bays to help absorb that sudden surge of power." Sure enough, a glance over the handrails around the exhaust ports nearby revealed lines of high-pressure delivery nozzles around the sides of each rectangular tunnel. These handrails, incidentally, are removed before launch, as are the numerous other 'delicate' items — pipes and vents and so on — that would otherwise come to grief in the holocaust of lift-off.

The other items of interest that caught my eye whilst we were walking around on top of the Mobile Launch Platform were the so-called 'hold-down posts'. Eight of these five-foot tall cast iron posts are attached to the lower extremities of the SRBs — four either side — to support the entire weight of the fully-laden Space Shuttle, and explosive bolts detonate, a fraction of a second before lift-off to release the vehicle from gravity's firm grip.

It was not without some trepidation that I was led 'below decks' into the very bowels of the launch platform. The urge to draw comparisons with the interior of a battleship or a submarine is irresistable. There's piping, electrical racks and ventilation shafts all over the place. It was surprising to see so much instrumentation inside the various compartments within the launch platform. Much of this ground support equipment is used for servicing the Orbiter when it's out on the pad, and there's apparently a lot of similar test equipment under the fixed concrete launch pads themselves. The hum of generators is a constant background noise. This submarine-like approach is not without purpose. The whole interior of the launch platform must be purged with gaseous nitrogen before launch to reduce the risk of a fire. When the heavy steel doors are tightly locked, each compartment is environmentally sealed to maintain the integrity of this purge. The entire interior is airtight; a sobering thought when you're walking around inside.

This type of purging operation is standard NASA practice at various stages in the Shuttle's launch preparations. But safety procedures are always treated with the utmost respect; no-one has forgotten the incident when two experienced Rockwell personnel lost their lives during a similar purging operation just four weeks before the Shuttle's maiden flight.

On that fateful day in March, 1981, a test was being conducted out on Pad 39A. The unoccupied aft end of *Columbia* was being purged with gaseous nitrogen, inevitably creating a low-oxygen environment. The 'All-Clear' signal to go back to work came as a result of a momentary breakdown in communications between the safety officers on duty at the time. There was still a lot of gaseous nitrogen in the compartment when five technicians entered it — and started dropping like flies.

Although they were rushed straight to hospital, two of them never recovered.

To prevent a repetition, there are monitoring stations throughout the interior of the Mobile Launch Platforms to warn the personnel working inside if a low-oxygen environment is imminent. Orange lights flash and an extremely noisy alarm sounds until the hazard is past. George Diller; "We've had a couple of occasions since the accident, both in the OPF and in here, when that system detected a low-oxygen situation and set off the alarm system, sending everyone running out just as fast as their legs would carry them. So it *does* work!"

I noticed a number of shower units located in various compartments of the Mobile Launch Platform, and enquired as to their purpose. George told me that they were for personnel who are unfortunate enough to come into contact with the hypergolic fuels which are used as an oxidiser by some of the Orbiter's engines. "Hypergolics are the single biggest hazard we have in here. They're very toxic. It's the residual action which is extremely injurous to skin and eyes, so you must flush your eyes out for fifteen minutes after exposure. If the safety people arrive and you haven't been under that shower for fifteen minutes, then they'll make sure you don't leave until you have. You're straight in with your clothes on — you don't waste precious time taking them off".

I was told that two Mobile Launch Platforms may not be enough to get the Shuttle up into the fast turnaround mode

Astronauts prepare to enter the M113 escape vehicle during training exercises on the emergency slide-wire escape system. The heavily-armoured vehicle is used to drive the astronauts and ground support personnel to safety once they have climbed out of the slide-wire 'basket'. Space Shuttle's escape system is better (some would say more humane) than that employed for the Apollo missions. The Apollo system comprised only a single basket, so there was a problem; if you missed the basket you were out of luck!

that NASA are aiming for. George Diller; "The launch platform you saw outside will be modified for Shuttle operations shortly. We'll reconfigure the inside and cut two more holes in it. The tower can be taken apart section by section, two levels at a time, so that if a Shuttle-derived vehicle comes along, who knows, we might have a tower all ready for it. We store the hardware we remove over at the Industrial Area until we need it. That way we save precious dollars". I was impressed by this continuing process of adaptation. Far better to modify existing hardware than to adopt a "we've got to have a new one" approach. It seems important not only that NASA save precious funds, but that they be *seen* to be doing so. But some items *will* have to be built from scratch. "We figure that we're gonna need four Mobile Launch Platforms to sustain a rate of 18-20 launches a year. The two we're currently using can get us up into the 12-15 range, but beyond that it looks like the launch platforms are gonna be the low pole in the tent if we don't modify that third one. Otherwise, you're gonna have hardware standing around waiting to get on it".

After the tour, my over-riding feeling was one of admiration; not only for the degree of technical expertise I witnessed, but also for the organisational effort that is involved in keeping KSC at work. NASA is not alone in running this vast operations centre on a day-to-day basis, of course. A whole host of subcontractors perform a wide variety of support services to ensure that KSC is kept operating as near as possible to peak efficiency. To provide an example of the diversity of tasks these support contractors provide, it's worth taking a closer look at the responsibilities of just one of them — Pan American World Services, Inc., a division of the celebrated airline formed by Juan Tripp in 1927.

In the early years of the Space Shuttle programme, World Services' Engineering and Design Department played a major role in the design of modifications to the Apollo launch facilities at KSC to accommodate this revolutionary new reusable spacecraft. The construction costs involved were in excess of $15 million. Principal facilities requiring extensive structural, mechanical and electrical modifications were the VAB, the Launch Control Center (LCC) firing rooms, the Operations and Checkout building, and several hypergolic fuel checkout facilities. Once all these facilities became operational, Pan Am World Services consolidated their involvement at KSC by monitoring chemicals, and checking noise and radiation levels before and after Shuttles leave their launch pads. They also provide a team of medical personnel

(Left) The dawn of a new era in spaceflight. A few seconds after 7:00 on Sunday 12 April 1981, the first Shuttle blast off, with Young and Crippen aboard.

(Right) A gigantic cloud of smoke is one of the highlights of a Space Shuttle launch. It is caused by the vehicle's mighty solid rocket boosters.

and specially-trained disaster teams who stand ready on an 'around-the-clock' basis to handle any emergency which may arise during the Shuttle's launch or landing phases.

In addition to these diversified tasks, Pan Am World Services also help run the U.S. Air Force's Eastern Test Range — a 10,000-mile-long chain of tracking stations stretching from Cape Canaveral to the South Atlantic. By means of radar and telemetry and photography, these stations track and record the Shuttle during its ascent from KSC's Launch Complex 39, and provide similar facilities for test launches of the U.S. Navy's Trident C-4 nuclear ballistic missilies and the USAF's powerful Titan III launch vehicle. World Services employ over 2,300 people to help operate the Eastern Test Range, and provide all kinds of unglamorous but vital facilities. These include much more than just the operation and maintenance of various launch facilities; there are more immediate responsibilities. Like the multitude of tasks associated with keeping the staff of thirty military, civil service and civilian contractor organisations well fed, providing them with security and fire-fighting services, and seeing that over 1,600 buildings and other structures that they use are properly painted, heated and cooled, supplied with electricity, water, sanitation services and otherwise kept in optimum condition. The company also has to maintain twenty-two miles of railroad track and associated switches and yards, not to mention fifty-two miles of fences and canals, and one hundred miles of roads — and all the grounds in between them — in the twenty-three square-mile Cape Canaveral area. Not the sort of glamorous work that is usually associated with the Space Shuttle project — but *someone* has to do it!

I was flabbergasted by the sheer diversity of tasks underway at Kennedy Space Center and Cape Canaveral Air Force Station at any given point in time.

After seeing Shuttle checkout and maintenance procedures at first hand, my tour of Kennedy Space Center concluded with a look over the facilities associated with actually launching them. My guide was Rocky Raab of KSC's press department.

Driving down one of the service roads towards Pad 39A, we came across something that compelled us to stop. Of all the extraordinary sights that can greet a visitor to KSC, the massive Crawler-Transporter vehicle in motion must rank as one of the most awesome. Affectionately nicknamed "The Mighty Tortoise" by KSC's engineering workers, the Crawler performs a vital task. It picks up the Shuttle vehicles — already assembled on their giant Mobile Launch Platforms — and carries them from the colossal Vehicle Assembly Building out to the launch pad. Once there, it lowers its precious cargo into huge pedestals and makes its way back to a special parking lot.

We watched for a few minutes as the gigantic machine trundled along its purpose-built road. The Crawler itself is as wide as a twelve-lane highway. Rocky summed it up in a few words. "Unladen, it's simply a six million pound bulldozer".

There are actually two of these gargantuan vehicles in existence; both of them based at KSC. But one is currently 'resting' prior to being returned to service at a later stage in the Space Shuttle programme. They were built in 1964 for the purpose of moving Saturn Vs from the VAB to their launch pads, initially for the Apollo Moon flights and later for the Skylab and Apollo-Soyuz missions. For the 5½-year period between the Apollo-Soyuz flight in 1975 and the start of launch preparations for the Space Shuttle's maiden voyage, both Crawlers lay idle. But they were kept in tip-top condition with bi-weekly maintenance and an extensive modernisation programme. The Crawlers had been designed in the early 'sixties, and their control systems reflected their age. Rocky Raab; "That was fifteen years ago, and some of the things that went into them were fifteen years old *then.* To prepare them for Shuttle operations, both Crawlers were updated from vacuum tube to solid-state technology. Modifications were also made to increase their reliability. There's an automatic shutdown mode should anything start going wrong mechanically, and new computer technology allows a self-test capacity. It does most of its own trouble-shooting; tells the technicians what's wrong, what's about to go wrong, its general state of health".

Once the Crawler came to a standstill, we were able to have a closer look over it. Both machines were built and modified to an identical specification. There are eight sets of tracks — two in each corner — and each set of tracks is composed of fifty-seven segments weighing one ton apiece. That totals up to over 450 tons in track segments alone! Power is provided by two, 2,750-horsepower diesel engines driving four 1,000-kilowatt generators, which provide electrical current to the vehicle's sixteen traction motors. Fully laden, the Crawler wouldn't win any prizes for fuel economy. With the Mobile Launch Platform and Space Shuttle vehicle on top — a total weight of 11,000,000 pounds (4,989,500 kilograms) — and travelling at a mere one mile per hour, it guzzles fuel at the rate of one gallon every twenty feet. It's somewhat ironic that the Shuttle must begin its 17,500mph journey into space with this snail's-pace crawl out to the launch pad.

For all its apparent brutishness, the Crawler-Transporter is a formidable piece of precision engineering. It has a special mechanism that keeps the Mobile Launch Platform absolutely level when negotiating the five-degree incline to the launch pad at the end of its journey. In addition, it is capable of not a little delicacy. To pick up and set down the Mobile Launch Platform on its huge 22-foot high pedestals at either end of its journey, the Crawler must manouver to within two inches of the platform's support points.

The cost in human lives, national and corporate reputation and hard dollars that would result if the Crawler failed to perform its task correctly demands that every possible precaution be taken on every single journey. For this reason, the crawlerway upon which the Shuttle-bearing vehicle must travel is also of prime importance. This roadway covers 3.5 miles (5.6 kilometres) in its link with the VAB and Pad 39A, and 4.25 miles (6.8 kilometres) from the VAB to Pad 39B, and is as broad as an eight-lane turnpike. It consists of two 12-metre (40-foot) wide lanes, upon which the Crawler's mighty tracks run with a relentless crunching noise on deep layers of river gravel. These two lanes are separated by a 15-metre (50-foot) wide median strip, giving the appearance of there being *two* crawlerways. This is of

A collaborative development by NASA and IBM, the Launch Processing System (LPS) is the highly-automated, computer-controlled overseer of the entire launch operation.

Approximately 300,000 gallons of water are dumped onto the launch pad in a 30-second period by these so-called 'rainbirds' to cushion the Shuttle vehicle from damage by acoustic energy generated mainly by the mighty solid rocket boosters. Acoustic suppression had to be stepped up following damage caused to the launch platform during the first lift-off.

course not the case; the Crawler occupies the entire strip and has driving cabs at either end, neatly alleviating the need to turn the gigantic machine around at either end. In this fashion, the Crawler vehicles have spent their entire 17-year lives on this one stretch of road, and will continue to do so until well into the 1990s, and quite possibly beyond that.

Despite their gargantuan size, the Crawlers have a character all their own; a fact confirmed by one particularly odd distinction they share. In 1977, both vehicles were declared National Historic Mechanical Landmarks by the American Society of Mechanical Engineers! There's another rather pleasing footnote to the Crawler story. In 1969, when Donald Buchanan — KSC's chief of engineering — travelled to London to accept the Diamond Jubilee trophy of the Royal Automobile Club for "The outstanding contribution in the field of automotive transport", he carried with him a small scale model of the Crawler vehicle. On seeing the model in Buchanan's suitcase, the customs officer on duty at the airport — obviously unaware of the vast proportions of the real-life Crawler — earnestly enquired if he was a travelling salesman selling farm machinery!

Driving away from the Crawler parking lot along the east side of the VAB, we cast an eye toward the Launch Control Center, where the so-called 'firing rooms' are. It is here that launch operation activities are focussed each time a Shuttle gets airborne. There are four firing rooms in the Launch Control Center, one on each end of the long, low building and two in the middle. Two have been modified for Space Shuttle operations (one each for Pads 39A and 39B), one lies dormant, while another is being modified not for a third launch pad, but for security requirements when future Department of Defense payloads are launched aboard Space Shuttle from KSC. Raab: "Those launches will be conducted by our people, but the DoD want to control access — not only person-wise, but also electromagnetic wise — in and out of that room. They want it electronically secure for their payloads. The checkout consoles and the computers in there will all be the same as those already in use, but the payload monitoring equipment and procedures will be different, to suit military payloads".

When DoD payloads are flown, ground controllers often talk in code to the astronauts in orbit, to ensure that security is upheld. Any enquiries as to the nature of DoD payloads being flown aboard the Space Shuttle meet with a firm "No comment".

A major goal of the Space Shuttle programme is to provide routine, economical access to space by frequent launches of a small fleet of reusable Orbiters at a cost well below that afforded by the expendable vehicles used during all previous programmes. KSC alone may well conduct up to forty launches a year, and the USAF will be launching Shuttles from facilities at Vandenberg Air Force Base in California. In order to be economical, such a high launch rate naturally requires a short, inexpensively-performed vehicle checkout and assembly operation, and this is accomplished by means of the Launch Processing System, or LPS. A collaborative development by NASA and IBM, the LPS is the highly-automated, computer-controlled overseer of the entire launch operation and is one of the most important aspects of KSC's drive for speed and economy. It provides a highly streamlined and efficient means of performing what was previously a slow, expensive and labour-intensive process, and it greatly increases the performance capability of the test engineers by enabling each of them to do the work of several people employed on previous programmes.

The main key to frequent and economical Shuttle launches is a short turnaround time; the length of time an Orbiter spends on the ground between missions. By holding the turnaround time to just a few weeks, a small fleet of Orbiters can satisfy the launch requirements of both KSC and Vandenberg. By comparison, the turnaround time of several months needed to assemble and check out the old manned

expendable vehicles such as Apollo would require a fleet of twelve vastly-expensive Orbiters, plus very high initial costs for the construction of new facilities and huge continuing expenditures for operating manpower.

To meet the many requirements for a short turnaround time, and significantly reduce the number of operating personnel involved, the entire launch processing operation for the Space Shuttle has been automated with the aid of digital computers. This includes automation all the way from the Shuttle vehicle itself right up to the test engineer sitting at his console. The LPS performs the bulk of the checkout work for the Orbiter vehicle, and much of that for the external tank and the solid rocket boosters. Test engineers monitor the LPS through its Checkout, Control and Monitor Subsystem from the firing rooms.

A firing room is 'dedicated' to a Shuttle from the moment an Orbiter arrives in the Orbiter Processing Facility and monitors and controls it through the multitude of assembly and checkout steps, all the way to launch. There are two working bays in the Orbiter Processing Facility, two stations in the VAB, two firing rooms (not counting the third 'secured' firing room being prepared for USAF Shuttle launches), two Crawlers and two launch pads, so that two or more Shuttles can be 'in process' simultaneously.

In addition to its checkout function, the LPS is designed to serve as the central reporting point for a multitude of associated activities. These include receiving and having available for display such information as status reports on work conducted in other buildings where Shuttle components are taken, as well as schedules and work control, and countdown and launch operations. In short, the LPS and its operators are the heart of what NASA hope will be a thoroughly economical checkout and launch system.

During assembly and checkout, the LPS 'talks' with the solid rocket boosters, external tank, Space Shuttle main engines, and the many complex systems aboard the Orbiter vehicle itself. It achieves this through extensions called Hardware Interface Modules, or HIMs. These are located in areas such as the Orbiter Processing Facility, the VAB high bays, the Hypergolic Maintenance Facility where some of the Shuttle's highly toxic propellants are stored, in KSC's extensive industrial area, out on the launch pads, in the spacecraft checkout buildings, and at various other sites supporting Shuttle maintenance and prelaunch checkout. It's an extremely complicated system, but — so far, at least — it works well. The acid test will come when launch frequency really hots up in the middle of the decade.

The 'human' side of the LPS became immediately apparent when Rocky and I walked into one of the firing rooms some weeks later, just before launch. Dozens of operators were sitting at their consoles; each subsystem operator position in a firing room has its own keyboard and visual display unit, and all consoles (they're arranged in semi-circles of six) are orchestrated to work together on major tasks through an 'integration console' at the front of the room. Each console can perform several tests or procedures simultaneously to save yet more precious time.

The small computer with each console has an on-line disc storage capacity of five million words, to hold all the test procedures to be conducted by the operator and his assistants. This system allows the checkout and launch functions of each console to be changed if necessary, either by means of reloading with data from the Master Console, or by physically moving discs from one console to another, providing the necessary flexibility and back-up capacity. In addition, the independance of each console permits the various systems to be tested in parallel, saving a great deal of time. There are about 100 parallel mini-computers and microprocessors in the entire LPS, about forty minicomputers being used during a normal launch operation.

The number of personnel on duty in a firing room at launch time is less than half of the 450 figure required to launch the Apollo vehicles. This reduction in manning levels is due in no small measure to the ability of the LPS to monitor literally thousands of measurements on the vehicle and ground support equipment simultaneously, compare them to predefined tolerance levels, and display only those values that are out of tolerance. The monitor feeds outputs to the console screens to display, in a variety of colours, those conditions that must be evaluated by the test engineer. A flashing red light, for example, might indicate a high-pressure condition which must be corrected immediately.

In many cases the LPS computers will automatically react to the beyond-tolerance conditions and perform safing or other related functions without test engineer intervention. All relevant data is stored for future reference.

A typical example of the LPS at work can be shown by following the liquid oxygen tank loading sequence when a Shuttle is on the pad. Only one console — numbered C-9 — is required. Its task is to transfer some 140,000 gallons (592,000 litres) of the supercold liquid (-181 degrees C, or - 295 degrees F) one-third of a mile from the liquid oxygen storage tank at the edge of the launch pad to the Shuttle vehicle's giant external tank.

First, the C-9 console operator performs several programmes to verify that the system is ready to begin the fill operation. These programmes establish

(Left) STS-8. *Challenger* casts a brilliant glow across KSC's marshy landscape.

The ability to make launches by night greatly increases the Shuttle's operational scope.

that 1) All tolerance limits are set to their standby conditions; 2) All system measurements are being reported; and 3) All mechanical valves are cycled to determine their readiness to operate. The LPS does all this without operator intervention, finishing — unless unusual conditions occur — in about ten minutes. When the verifications are complete, the C-9 operator awaits a "GO" signal from the lead test conductor. When it comes he pushes a single button marked "FILL". The liquid oxygen loading operation begins, and continues automatically until completion. It includes these major steps, listed in the sequence in which they occur:-

1: A ten-minute chilldown of the storage area and pump. (Liquid oxygen flashes into gaseous form if it contacts surfaces warmer than –181 degrees C). All pipes, pumps, tanks, etc., have to be prechilled to prevent an excess amount of gaseous oxygen from forming.

2: A chilldown of the Orbiter's main engines, through which the liquid passes on its way to the external tank, and the liquid oxygen tank itself.

3: A slow fill of the liquid oxygen tank until it is two percent full.

4: A much faster fill of the oxygen tank until it is 93 percent full.

5: A slower topping up of the tank to 100 percent full.

6: The constant replenishment of the oxygen boiled off and bled off in gaseous form until about nine minutes before launch.

7: At ignition minus three minutes the liquid oxygen tank is sealed, pressurised and ready for launch.

Some two hundred computer programmes are required to complete all these phases of action. They operate a primary pump, primary fill valve or secondary fill valve, etc., throughout a complex piping system. While these programmes are in process some 150 measurements are constantly monitored to be certain all temperatures, pressures, etc., are maintained within carefully-prescribed limits. If a condition is detected which requires immediate corrective action, the programme takes that action and notifies the operator. Less immediate problems are called to the operator's attention for his consideration. The operator has the power to alter the sequence of events, or take over control, in the unlikely event that he should think this necessary.

The loading of liquid oxygen into the Shuttle's external tank is only one of hundreds of equally complicated operations performed automatically by the LPS, while operating under stringent safety and performance requirements.

Approaching Pad 39A on our journey along the crawlerway, the two giant insulated spheres which supply propellant for the Shuttle's main engines (SSMEs) came into view. They are situated at opposite sides of the pad for safety. Those on the northwest corner of each pad have a capacity of 3,406,500 litres (900,000 gallons) of liquid oxygen, which the SSMEs use as an oxidiser. Those on the northeast corner can hold 3,218,250 litres (850,000 gallons) of liquid hydrogen, the SSME's fuel. One thing the two propellants have in common is that they are both cryogenic, a word of Greek derivation — kryos — meaning "ice cold". Ice is warm in comparison to these two cryogenics — liquid oxygen boils off as a gas at temperatures in excess of –181 degrees C (–298 degrees F) and liquid hydrogen must be kept at temperatures at or below –253 degrees C (–423 degrees F) if it is to remain in the liquid state. So the huge spherical tanks in which these propellants are stored at the launch pad can be thought of as a king-sized thermos flask.

In gaseous form, oxygen and hydrogen have such low densities that extremely large tanks would be required to store them in the spacecraft itself. But cooling and compressing them into liquids vastly increases their density, making it possible to store huge quantities. The distressing tendency of cryogenic liquids to return to the gaseous form unless kept super-cool makes them difficult to store over long periods of time, and hence less than satisfactory as propellants for military rockets, which must be kept launch-ready for months at a time. But the high efficiency of the liquid oxygen/liquid hydrogen combination — liquid hydrogen has about forty percent more 'bounce to the ounce' than other rocket fuels and is also very light — makes the low temperature problem worth coping with in applications where reaction time and propellant storability are not too critical. The ability to use hydrogen means that a given mission can be accomplished with a smaller quantity of

The solid rocket booster recovery ship *UTC Freedom* ties up overnight at the Trident Submarine Basin at Port Canaveral following retrieval of a spent SRB casing. *UTC Liberty* will later take the other SRB to Cape Canaveral AFS.

propellants — and hence a smaller launch vehicle — or, alternatively, a mission can be accomplished with a larger payload than is possible with the same mass of conventional propellants.

At the end of the long crawlerway we drove up the five-degree slope onto launch pad level; a concrete plateau 45 feet above the surrounding area. In the distance we could see Pad 39B, which is being prepared to an identical specification to enter service in early 1986 to meet the increasing frequency of Space Shuttle launches. A Shuttle launch gantry comprises two principal components; the fixed service structure and the rotating service structure. The fixed service structure is a square cross-section steel tower that provides access to the Orbiter at many different levels. It is painted orange and incorporates several sections of the Saturn V launch towers removed from the old Apollo platforms. The hammerhead crane on top is still basically identical to the original Apollo specification.

But to this basic orange-painted structure are numerous grey-painted additions. These are the brand new sections incorporated for Space Shuttle operations. By far the most significant addition is the rotating service structure; a bewildering mass of struts and girders which pivots about a column of massive hinges on the fixed service structure to completely envelope the Shuttle vehicle for the purposes of access during preflight preparations. Only at a fairly late stage in the countdown does this structure swing clear through 120 degrees.

There was plenty to look at from our vantage point at the edge of the launch pad. The two 'white rooms' were clearly visible. These environmentally 'clean' compartments form a seal with the two parts of the Orbiter which open to allow ingress of personnel and equipment; namely the cockpit and the payload bay. The flight crew's 'white room' is located at the end of an extending arm attached to the fixed service structure, and only retracts from the Orbiter's crew access hatch two minutes before launch. There is enough room inside to accommodate six people at any given time.

The payload bay 'white room' is mounted on the rotating service structure and is rectangular to match the shape of the cargo hold opening where payloads which must remain in a vertical position are loaded aboard. Around the outside of the payload bay 'white room' doors is a fabric dust seal which is pressed against the upper surfaces of the Orbiter to maintain the integrity of the environment within. These doors must be scrupulously cleaned before they are opened to ensure that no contamination occurs. Payload transfer is an extremely involved operation.

Other umbilical arms were visible on the fixed service structure, one of which provides access to the intertank area of the external tank. This allows any defective components within to be replaced out at the pad, instead of taking the whole vehicle back to the VAB on the Crawler-Transporter. Apparently, technicians also enter the intertank area just before launch to install flight batteries and safing and arming plugs for the Shuttle's self-destruct system. More about this later.

It was interesting to note that the rotating service structure, though it pivots about huge hinge assemblies, is supported from beneath by a massive trolley mounted on rails. This comes complete with its own driver's cab. Rocky broke in again; "It's a short trip, but a slow one!"

Many other interesting items caught my eye out on the launch pad. One was the gigantic wedge-shaped flame deflector device which diverts the white-hot exhaust columns from the solid rocket boosters into the 42-foot deep flame trench beneath the launch pad, from whence it pours out from the sides and forms great white billowing clouds. The blast deflection device is mounted on rails so that it can be moved back to allow the Crawler to straddle the 58-foot wide flame trench with its enormous tracks.

I asked Rocky about damage caused to the launch pad during *Columbia's* blast-off on the Shuttle's maiden voyage. Some newspaper reports had intimated that NASA had underestimated the violence of the Shuttle's departure and that damage to the pad had been considerable. "The thing we hadn't anticipated was the crabwise tracking motion the Shuttle made towards the fixed service structure immediately after lift off. It missed the gantry by only three feet or so. But, in actual fact, the damage was less than that sustained during the early Apollo launches. We've had progressively less damage from each Shuttle launch as we've gradually learned what had to be done to protect the most vulnerable components on the fixed service structure and Mobile Launch Platform. The early damage was caused not so much from the heat of the rockets, but from the blast effect. We have handrails, loudspeakers, pipes, wires and lights all over the place, and many of these items were literally torn loose and blown into other things. So what we've done is to relocate the more vulnerable items away from the blast, and beef-up other bits and pieces to make them more resistant to the launch environment. Other items are removed altogether before launch. Since those modifications were completed we've had no damage whatsoever".

I asked what effect the blast effect had on birds flying in the vicinity of the launch pad. Cape Canaveral is, after all, an official wildlife refuge. "At the first sound from the Shuttle engines the birds fly off, but some do get caught. Again, it's not the heat that kills them, it's the acoustic levels. We've registered levels of 150-160 decibels out at the pad perimeter, and that's far beyond what you need to kill any living thing".

Next, Rocky pointed out the slide wires running from the fixed service structure down to the west side of the pad. There are five 1,200-foot-long cables and five 'baskets' for the Shuttle's (up to) seven

Department of Defense rescue personnel remove astronaut-suited dummies following a simulated crash landing of an Orbiter near KSC. Injured astronauts would be whisked away to a special medical facility at KSC following an aborted launch or landing.

crewmembers and the eight-man close-out crew to use in the event of an emergency just prior to launch. Each 'basket' is capable of accommodating three people, and they can go down separately or all together. At the bottom there is an arresting net, a system of decelerating chains and a concrete bunker. The Shuttle's slide wire system is better — some would say more humane — than that employed for the Apollo missions. The Apollo system comprised a single 'basket' which could hold nine people, but as Rocky pointed out sardonically, "There was a problem. If you missed the basket you were out of luck".

Turning into the prevailing wind from the east I noticed what appeared to be a cloud of fog blowing inland from the beach about half a mile away. Rocky told me this was actually a haze of salt particles which continually comes in from the ocean. This haze is of course highly corrosive, so the launch facilites have to be repainted endlessly whether launches are taking place not. Raab: "It's a constant fight. We use a yellow-coloured zinc chromate anti-rust primer, then cover it with either the orange or the grey top coat, which is also anti-rust. Then, in a few weeks time, we'll have to chip it off and start all over again, because *nothing* is corrosion-proof out here".

Before we left the pad, Rocky drew my attention to the tall lightning conductor mast atop the fixed service structure, the tip of which stands 347 feet above the launch

This spectacular photograph was taken with a hand-held camera by mission specialist astronaut Kathryn Sullivan from the rear seat of a T-38 chase aircraft.

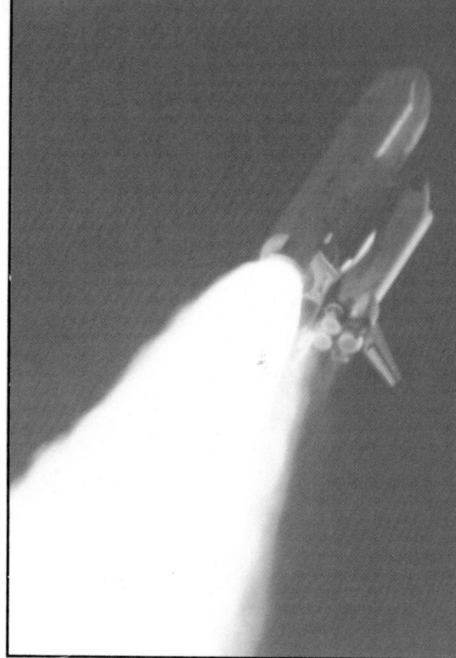

The Shuttle's solid rocket boosters spew out white-hot columns of flame for the first two minutes of each mission.

pad. The mast itself is actually an insulator. It is the two long cables sloping down from it to opposite ends of the pad perimeter that are the lightning conductors. Due to these cables, lightning from the east would — electrically speaking — 'see' the entire launch pad as a cone and would strike the cables, rather than the spacecraft and launch facilities contained within this 'cone of protection'.

As we passed the guardroom at the main gate to Pad 39A on our way out, we had to pull off to one side to let a fleet of tanker trucks through. They were delivering liquid oxygen to the LOX storage tank in preparation for the next launch. Rocky said that each mission required about fifty tanker-loads of liquid oxygen alone. Apparently, the insulation on the spherical storage tanks is so good that propellant can be stored almost indefinitely.

To watch any rocket launch is an astonishing experience, but Space Shuttle launches rank amongst the most spectacular of all. Representing the British aviation magazine *Air Extra* at the Shuttle's maiden launch in April 1981, I was privileged to join the 3,800 or so members of the world's press that had congregated at Kennedy Space Center to witness what promised to be a truly unforgettable event. NASA told us this was the largest gathering of press people in history, and who were we to argue; the Shuttle programme had attracted a degree of public interest that was without precedent in the history of spaceflight. A press tour the day before — during which we had been permitted to get quite close to the Shuttle on Pad 39A — had certainly whetted our appetites for the launch. But it was not to be. A 'time-skew' malfunction in the Shuttle system's backup computer stopped the countdown at T minus nine minutes — so near yet so far.

After the keen disappointment of Friday's aborted launch, the tension and excitement in the hours prior to the second attempt on Sunday 12 April was noticeably less intense than before. True, over one million people had packed themselves onto

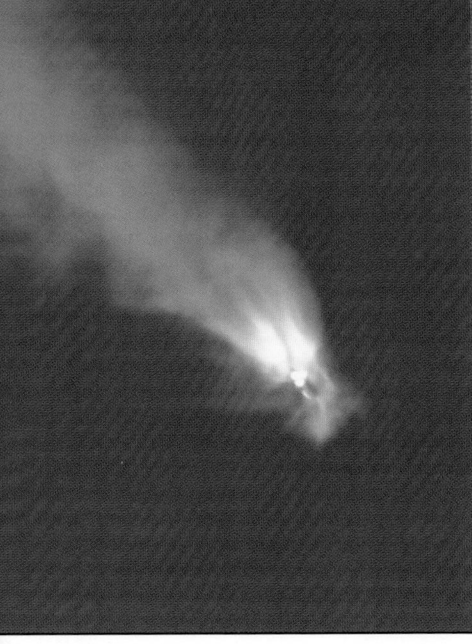
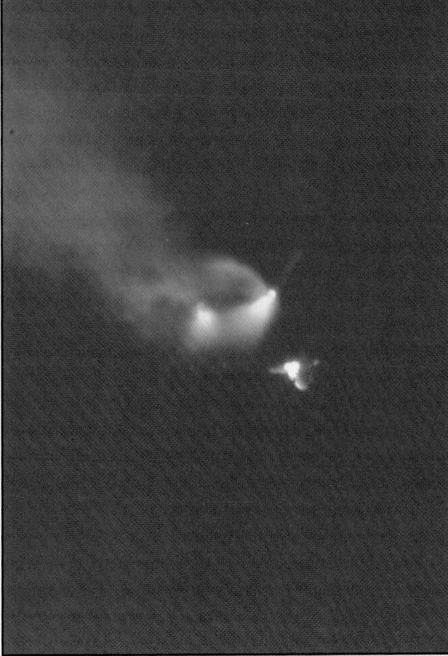
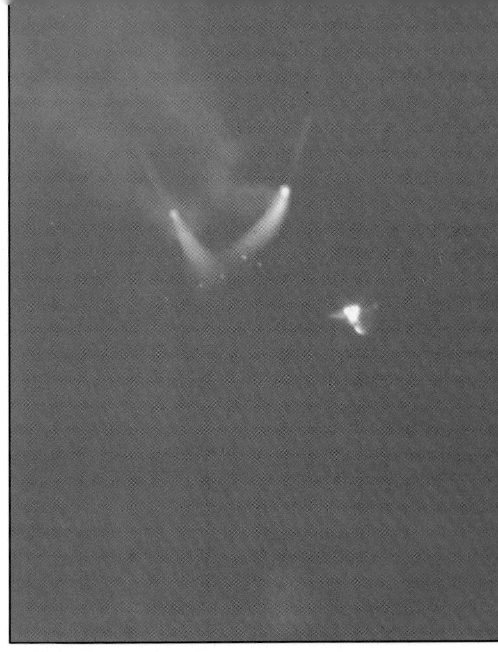

Approximately two minutes after launch, clusters of small rockets at either end of each solid rocket booster (SRB) push them clear. Their own momentum causes them to coast upward in a ballistic trajectory for another seventy seconds — to an altitude eleven miles higher than that at which they were released! They attain a speed of almost 3,000mph before parachutes lower them into the ocean.

every conceivable vantage point in the area, bringing traffic on roads for miles around to a standstill and congesting the nearby beaches. A galaxy of stars stood expectant in the VIP's encolsure at KSC; Lisa Minnelli, Clint Eastwood, Steven Spielberg, *Star Wars* director George Lucas, Pat Boone, John Denver. Ex-astronauts and dozens of U.S. Senators and foreign ambassadors to the USA rubbed shoulders with over 80,000 invited guests within KSC's boundaries. Some people even sat in boats moored on the Banana River to get an unrestricted view. But it seemed as if no-one could believe that the big event was really going to take place. I for one was anticipating the announcement of yet another last-minute technical hitch with every passing second.

It was only when the long countdown resumed after the scheduled ten-minute hold at T minus nine minutes, when a spontaneous burst of appreciation erupted from the spectators' enclosures, that the sheer presence of the moment gained its former strength — increased, if anything, now that we were all in uncharted territory once more. In the airspace above KSC, there was plenty of activity. During the hours of darkness, when the launch site was ablaze the world's most intense floodlight system, an aircraft — its undersurface bedecked with a multitude of tiny white lights — had spelled out the words 'OMNI MAGAZINE WISHES GOD-SPEED TO THE CREW OF COLUMBIA'. Now, with daybreak dawning C-130 weather surveillance aircraft and aircraft equipped with high-speed cameras wheeled and turned expectantly, while the CH-53 rescue helicopters sped out to their prearranged holding points in preparation for any emergency which might arise.

The launch-controller's voice caught everyone by surprise as it echoed those magical words across the Cape . . . "nine, eight, seven, six, five, four . . .," then belching clouds of smoke and steam spewed out from the flame trenches beneath *Columbia's* main engines and engulfed the launch platform. Nothing else seemed to happen for a few agonising seconds — a terrible vision of disaster flashed through my mind — then a searing white-hot spectre of flame pierced the dense, swirling ocean of grey as the solid rocket boosters ignited. *Columbia* stood firm momentarily as her five ascent engines realised their full potential, when explosive charges sheared the hold-down bolts that shackled her to the launch platform. Then she started to move, so slowly at first that you felt certain she was going to topple over.

Resolute now in her determination to gain altitude, *Columbia* cleared the tall lightning mast atop the fixed service structure, imperceptibly gathering momentum as she clawed her way up into the sky. The discharge from the solid boosters formed two brilliant white columns beneath the Shuttle as she rose, then a computer-controlled 'roll program' pirouetted the giant craft round and sent her out on a graceful curving arc over the Atlantic Ocean.

It took a few seconds for the sound of *Columbia's* rockets to cover the 3½ miles that separated the press enclosure from Pad 39A. It first reached us as a deep rumbling noise, fearful, almost deafening. But as the spacecraft gained height this menacing low tone became a mere backdrop to a phenomenon I was totally unprepared for as a first-time launch witness — a fierce crackling noise of awesome itensity. It was the sound of raw energy, like static on a radio but infinitely louder; as if the sky were a gigantic sheet of canvas being ripped apart. It manifested itself physically by flapping our clothes, blowing our hair and shaking the ground for miles around. For fifteen seconds or more, everything reverberated to the

One of the heavy-duty parachutes used by an expended SRB is 'tented' following an ocean splashdown. This procedure lifts and opens the fabric to allow meticulous inspection and permit every inch to be checked for tears and debris. Afterwards, the big ribbon canopy is put through a thorough wash and dry process.

relentless hammering, pulsating rhythm of countless shockwaves — an extraordinary phenomenon which has to be experienced at first hand to be fully appreciated. Then, slowly, the sound and vibration began to diminish as *Columbia* curved away into the distance, hidden now by the two solid pillars of white smoke that hung in the still Florida air long after the craft had departed from view.

A brief 'throttling back' period is calculated to occur at the most critical part of the ascent — the so-called 'Max Q' period, which begins fifty-three seconds into the mission — when the Shuttle vehicle is being subjected to the most severe aerodynamic and acceleration-imposed loadings. At this point, when the vehicle needs a respite from the forces acting upon it, the output of the Shuttle's five ascent engines is reduced.

Two minutes and four seconds after launch, the solid rocket boosters (SRBs) part company. Clusters of tiny rockets at either end push them clear. Once released, the SRBs do not immediately begin their parachute descent into the sea. In fact, their own momentum causes them to coast upward in a ballistic trajectory for another seventy seconds — to an altitude eleven miles higher than they were released! The boosters attain a maximum speed of approximately 4,650 km/hr. (2,890 mph) before being slowed down by atmospheric drag. It is four minutes after they parted company with the spacecraft before a series of parachutes of increasing diameters play their part in bringing the $20 million SRBs safely back to base.

The SRB parachutes are probably the strongest ever built. In view of the deployment speeds involved, this is not surprising. Their canopies are composed of concentric nylon ribbons spaced like a venetian blind; a construction that bestows a particularly high tensile strength. At an altitude of 4,694 metres (15,400 feet), a barometric switch actuates three small thrusters near the top of each SRB to eject their small nose caps. As the nose cap moves away, a pilot parachute is deployed to lower it slowly downwards. At this point, cutters release a drogue 'chute to begin the process of decelerating the SRB itself. This

One of the specially-designed and manufactured parachutes used to lower the Shuttle's expended solid rocket booster casings into the ocean is recovered by *UTC Freedom.* After the parachutes are reeled in they are taken for reprocessing at KSC.

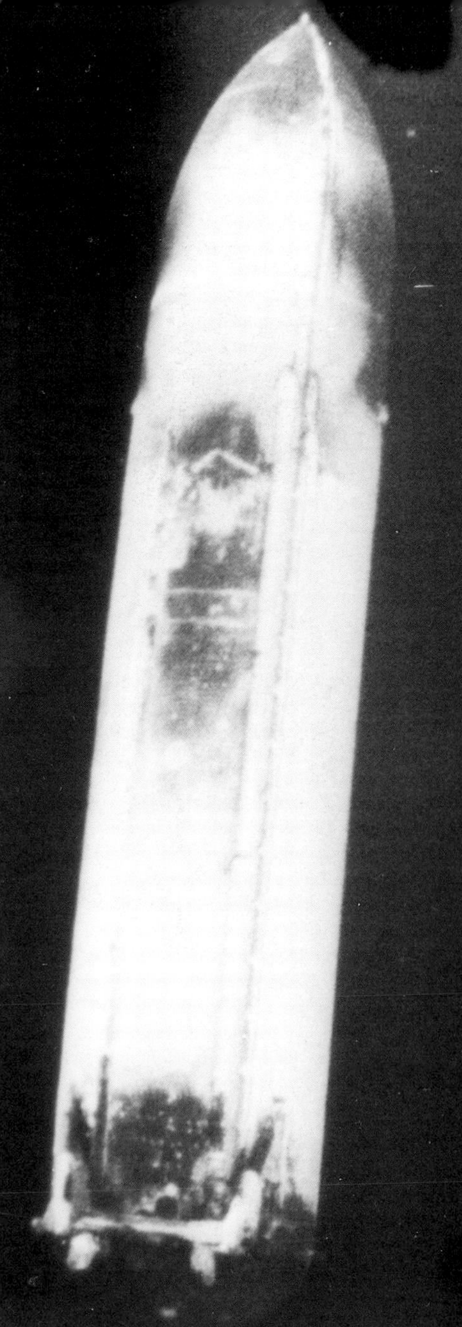

(Above) Separation of the Shuttle's external tank, photographed by a motion picture camera in the Orbiter. *(Below)* SRB recovery. Note charring from the cluster of four small separation rockets.

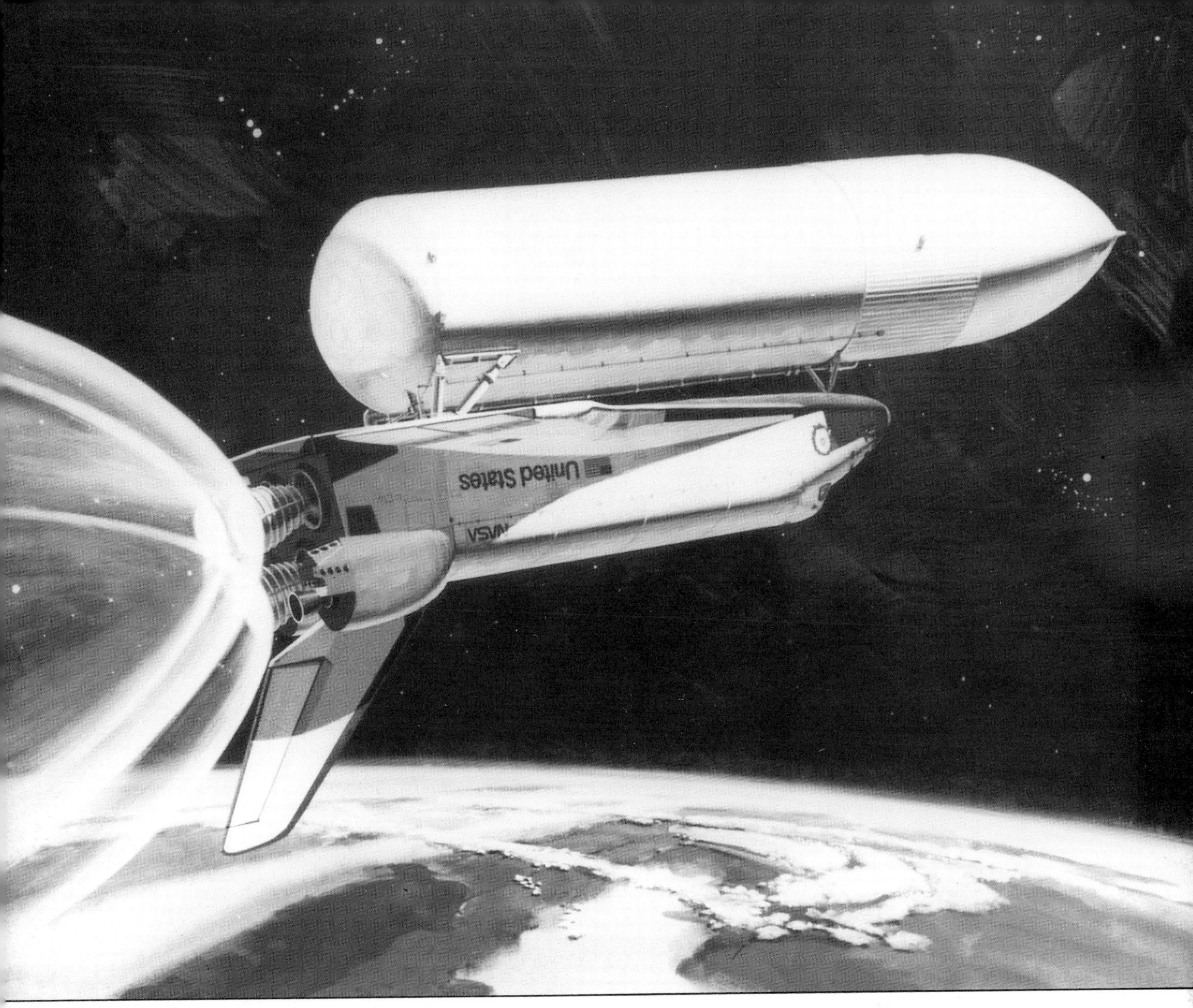

On its way into orbit the Orbiter carries its external tank piggy-back style until the propellants within have been expended. Explosive charges release the giant tank, which disintegrates into tiny charred fragments due to friction with the denser levels of the Earth's atmosphere. The fragments fall into the India Ocean, well away from shipping lanes.

initally inflates to sixty percent of its full size until the speed has been sufficiently eroded for a reefing line to be cut, providing eighty percent inflation.

At an altitude of 2,835 metres (9,300 feet) a second reefing line is cut to allow full parachute inflation. Another critical stage is reached when a second signal from the barometric switch activates an explosive charge to separate the top section of the SRB — the 'forward frustrum' to give its correct name — which continues its descent under the drogue 'chute while a cluster of three main parachutes sprout from the top of the SRB proper and lower it on an enormous 52-metre (172-foot) line. These main 'chutes undergo a double dereefing sequence which reduces the final impact speed of the SRB to 111km/hr (69mph). Upon splashdown, an impact switch sets off yet another explosive charge, this time to release and separate the main 'chutes and allow them to drop into the water nearby.

Lift-off from the launch pad took place a mere seven minutes ago.

Bobbing slowly up and down in the water in a vertical position, the SRBs have to sizzle away on their own for a short time until the two retrieval ships awaiting them about 160 miles (258 kilometres) down-range position themselves alongside. The boosters can splashdown anywhere in an 11 x 17-kilometre (7 x 10-mile) area. The retrieval ships are the *UTC Liberty* and the *UTC Freedom* (UTC stands for United Technologies Corporation, parent company of United Space Boosters Inc. who operate the ships). Both vessels are leased by NASA, and are powered by twin 2,900 horsepower diesel engines.

The spent SRB casings and the frustrum-drogue combinations carry radio signalling devices that allow *Liberty* and *Freedom* to home in on their precise locations. The main parachutes are located by sonar and are the first items to be recovered if the booster casings appear to be in a stable condition. Their winch lines are wound onto three of the four large reels situated on the rear decking of each ship. Then the drogue parachutes attached to the frustrums are reeled in by the same method until they are close enough for the deck cranes to hoist them up out of the water.

Recovery of the SRB casings themselves is the final retrieval operation. This delicate task is carried out by means of two strange-looking dewatering units. Each one is basically a nozzle plug which can be manouvered underwater by remote control, linked to a generator and an air compressor. The nozzle plug is inserted into the end of the SRB casing and forces the seawater out with compressed air. As the casing is pumped out it rises up into a horizontal position on top of the water ready for towing. Depending on the weather conditions it can take anything from two to five days to get the spent SRBs pumped out and towed the 258 kilometres (160 miles) or so back to Port Canaveral, where they are pulled into the disassembly plant in Hangar AF on the eastern shore of the Banana River on Cape Canaveral AFS.

The first stop for the SRBs after arrival at the plant is the washout facility, where any foreign materials, including salt water residues, are removed with fresh water. Each SRB is then separated into its four constituent 'casting segments', loaded onto trailers and transported back to Morton Thiokol's Utah facility, where combustion residues and insulation are literally cut out by means of water jets operating at extremely high pressure. After this, each segment undergoes magnetic particle inspection to determine whether any cracks or defects have appeared. Next, the case segments are filled with oil and pressure-tested, followed by a second magnetic inspection to ensure that the test itself has not resulted in any structural damage.

Finally, the casings are reassembled into casting segments, repainted, rein-sulated and reloaded with propellant prior to the next flight. While all this has been going on, some parts of the nozzle, igniter hardware, and safety and arming devices have also been refurbished for reuse.

With the exception of the STS-4 mission, when both SRBs were lost in 600 fathoms as a result of a malfunction in the parachute deployment sequence, the Shuttle's solid boosters have proved themselves to be more than a match for the gruelling environment they have to operate in, although the twenty-flight 'lifetime' capability has yet to be demonstrated.

In the case of the STS-1, *Liberty* and *Freedom* located the floating SRBs easily, but their retrieval crews found it impossible to completely purge the casings of sea water, as slight damage had been sustained to the thrust-vector control system of one booster, while the other had suffered some delamination of its exit nozzle. Consequently, the SRBs were towed back to Port Canaveral while lying at an angle in the water and were finally brought ashore on the afternoon following the launch. We found out later that the damage sustained to the boosters was not so great as to preclude their reuse.

Meanwhile, out of sight of the naked eye, *Columbia* was actually climbing faster than planned. Eight minutes and forty-five seconds into the mission the main engines were turned off, and twelve seconds later the giant external tank fell clear. The tank began to disintegrate at an altitude 30,000 metres higher than anticipated, thus littering a wider area of the Indian Ocean with tiny particles of debris. Due to the random tumbling motion of the tank on its way back down into the atmosphere, its exact trajectory is in any case very difficult to predict. But the impact area is well away from any land masses and shipping lanes, so this is not of any real consequence.

The next major hurdle was cleared when two Orbital Manouvering Systems (OMS) 'burns' — at 10½ minutes and 44 minutes respectively — placed *Columbia* safely in Earth orbit.

One of the questions most frequently asked about the Space Shuttle's launch procedure is, "What if something goes wrong?". This question became even more relevant after STS-4, when *Columbia's* ejection seats — similar to those fitted in the Lockheed SR -71 'Blackbird' reconnaiss-ance aircraft — were rendered inoperable, soon to be removed altogether. *Challenger,* which entered service well after the first four proving flights, has never had ejection seats fitted, and neither have the other Orbiters.

There are actually three contingency

operations the crew can initiate if an emergency arises during ascent. The one they select depends on the altitude they have managed to gain beforehand.

Return to Launch Site (RTLS): This contingency could be required immediately after lift-off, but cannot be initiated until after SRB separation. (The external tank would be jettisoned soon after). The SRBs cannot be jettisoned until they are virtually spent, so RTLS would therefore be initiated between about 2 minutes 30 seconds and 4 minutes 27 seconds after lift-off and would result in a landing on KSC's Orbiter Landing Facility (or Vandenberg's, if that has been the launch site) about thirty minutes after lift-off.

Abort Once Around (AOA): This is available as an abort mode between about 3 minutes 50 seconds and forty minutes after lift-off, by which time a Shuttle launched from KSC has gained enough altitude to reach Northrup Strip on the White Sands Missile Range, New Mexico, ninety minutes after lift-off.

Abort to Orbit (ATO): Available as an abort mode between about four minutes and forty minutes after launch, if the nature of the emergency is such that an immediate return to Earth is not necessitated. By this stage in the ascent, the vehicle is virtually free of the Earth's gravitational pull. The ATO contingency is planned to result in one Earth orbit, followed by atmospheric re-entry to a landing at the launch site. If the re-entry flight path does not turn out exactly as planned, the crew have the option of diverting to Northrup Strip.

SAR (Search and Rescue) helicopter teams would search for and locate a ditched or crashed Orbiter and recover the crew in the event of a deviation from the spacecraft's preplanned descent path in the case of all three contingencies.

These contingencies should cover any event which is likely to occur during the early stages of a mission, but one little-publicised aspect of the launch procedure is the Shuttle's built-in self-destruct capability. There are a number of heavily-populated areas around KSC, such as the city of Titusville a couple of miles to the west, that could become prime targets for catastrophe should the spacecraft go off course during its ascent. In the highly unlikely event of this happening, the Range Safety Officer in charge of this aspect of the launch operation can press a button and cause the vehicle to explode into relatively harmless debris.

During the first four flights, when the crew were provided with ejection seats, they could have been warned of the impending self-destruct programme and get out as quickly as possible. But what happens now that the ejection seat option has gone? In this situation there is no means of escape for the crew if any emergency arises, so the Range Safety Officer could be confronted with an agonising decision. Does he kill up to seven helpless crewmembers, or does he jeopardise the population of an entire township? It's an interesting question, and one that — perhaps understandably — NASA have remained tight-lipped about thus far . . .

(Pictures at extreme left and above) At White Sands Missile Range, New Mexico, a full-scale mockup of the Shuttle Orbiter's cockpit area was attached to a rocket propelled sled and used for ejection seat tests. The astronaut mannequin separated from his seat, whereupon 'he' was lowered onto *terra firma* on his own parachute.

LIFE ABOARD THE SPACE SHUTTLE

In the vacuum of space, moving at some 27,300 kilometres per hour, a Space Shuttle Orbiter can whirl around the Earth equally well in any attitude; nose first (which 'looks right', like an aircraft), or tail first, or even broadside. Either the underside or the upper surfaces of the Orbiter may face the Earth. The astronauts never know which way 'up' they are travelling until they look out of the window. In weightlessness they feel no 'up' or 'down' in any case. An Orbiter on operational flights is positioned according to the needs of the payload it is carrying, but various attitudes cause problems of temperature on board. Early test flights registered the reactions of the Orbiter and its payload to the most extreme temperature differentials that would ever be encountered during later, operational flights. Temperatures may range from +200 degrees C on a surface irradiated by sunlight to −200 degrees C on any surface out of direct sunlight or shaded by another object.

The Orbiter was flown in a variety of attitudes for extended periods of time to create these extreme thermal conditions. With relatively small satellites, the resulting thermal strains can be checked in advance in test chambers on Earth. But such tests were not possible with the huge Orbiter vehicle. During the first five Shuttle flights, temperature gauges were embedded in many places over the fuselage and wings. On one particular mission, the Orbiter kept its nose pointing towards the Sun for sixty hours to demonstrate the application of a technique to maintain tight temperature control over the instruments in the spacecraft's cargo hold. The pallet-mounted experiment package remained entirely in the shade and bitter cold throughout this phase. Then, for 34 hours, the Orbiter turned its tail to the Sun and rolled as it circled the Earth. The under-surfaces of the craft constantly faced the Earth, again leaving the payload bay frigid.

Next, the attitude of the Orbiter was altered to roast the contents of the payload bay in the full intensity of solar energy for 26 hours. And, finally, the 'barbecue mode' was tested. In this manouver, the Orbiter rolls slowly to equalise the temperatures on all its surfaces for up to twelve hours at a time whenever required, but especially before the fiery plunge into the Earth's atmosphere; an action formally referred to as passive thermal control.

To change orbit or rendezvous with another spacecraft, the crew uses the Orbiter's OMS (Orbital Manouvering System) engines. These are located in two external pods on either side of the tailfin. Pitch, roll and yaw control are accomplished by means of the Reaction Control System (RCS), which has 38 primary thrusters and six smaller vernier thrusters for 'fine-tuning' manouvers. RCS nozzles are located adjacent to the OMS engines and in the Orbiter's nose section.

Both the OMS and the RCS use hypergolic propellants; fuels and oxidisers which ignite on contact and need no ignition system. This easy start/re-start capability is ideal for manouvering systems. Another advantage of hypergolics is that they do not have the extreme temperature requirements of the supercool cryogenic propellants used by the Space Shuttle's main engines. The OMS/RCS fuel is monomethylhydrazine (thankfully abbreviated to MMH) and the oxidiser is nitrogen tetroxide. MMH is a clear nitrogen/hydrogen compound with a distinctly fishy smell. It's a kissin' cousin of ammonia and is highly toxic. Nitrogen tetroxide, on the other hand, is best described as a super nitric acid. It has the sharp acrid smell of many acids and, like MMH, is very toxic. Accidental exposure to either of these fluids could ruin your whole day.

(Above) **Satellites are set spinning to stablise them, before springs push them clear. A series of Payload Assist Module 'burns' permit orbital insertion.**

(Right) **This almost abstract photo — taken from a satellite just launched from *Challenger's* cargo hold — shows the Orbiter travelling at 17,500mph!**

One of the first and most vital operations the crew must perform once the Orbiter is safely in orbit is to open the giant payload bay doors, because attached to the inside of these doors are huge silver radiators, whose 1,800 square feet dispel the excess heat created by the spacecraft's operating systems and crew compartment into the cold vacuum of space. Built by Vought Corporation, perhaps best known as manufacturers of warplanes but also responsible for developing the radiators which performed faultlessly on the Apollo trips to the Moon, the radiators are composed of eight contoured panels, each fifteen feet long by ten feet wide, composed of thin aluminium sheets over bonded aluminium honeycomb. The exposed surfaces of the panels are covered with mirror-bright, silver coated Teflon to reflect the intense sunlight

encountered in orbit, and to enhance the radiation of heat into space. The panels themselves contain over a mile of tubing through which Freon 21 coolant is circulated to carry heat away from the various subsystems. The number of panels used depends on the heat-rejection requirements of various payloads. For a payload needing to reject 21,500 BTUs, for example, only the forward six panels are used. But for more demanding payloads, such as the Spacelab orbiting laboratory, which requires nearly 30,000 BTUs of waste heat dissipation, all eight panels are brought into operation. The Shuttle's radiator system has approximately ten times the heat-rejection capability of the system employed on Apollo; 100,000 BTUs in all. A fuel control system regulates the flow of Freon 21 (chosen for its wide boiling-to-freezing range and well-defined freezing point of 211 degrees below zero F) to a steady temperature of 38 degrees F, adjusting almost continuously as the radiators pass from bright sunlight into the Earth's cold shadow.

To the Shuttle's astronaut crew-members, life in orbit will gradually become routine; all in a day's work. But to we lesser mortals, the experience would be mind-boggling. A sensational panorama lies beyond the cabin windows. The Earth below is dun-coloured, with fluffy swirls of white cloud and blue-green oceans. Snow clings like cobwebs to mountain ridges and sand dunes in the great deserts cast subtle shadows at certain times of the day. The naked eye in orbit can discern even finer detail; smoke from steam trains and the wakes of ships have been spotted on occasion, not to mention such minutiae as the Vehicle Assembly Building and the Orbiters' runway at KSC.

The idea that ordinary people would someday live and work in this amazing environment has fascinated serious scientists and engineers as well as science fiction fans. Space Shuttle is the first step in turning this dream into a reality. Launch and re-entry forces are only about 3G (three times normal gravity); well within the limits which can be tolerated by healthy people. The Apollo lift-offs, by comparison, exposed their crews to up to 6G. So getting people into space is becoming easier. But what about the conditions they will encounter when they get there? Orbiter living accommodations are relatively comfortable; they incorporate advances made through nearly two decades of experimental manned space missions and an even longer period of ground studies. The Orbiter's cabin air is cleaner than Earth's and hay fever sufferers would welcome its pollen-free atmosphere. Air pressure inside the Orbiter is the same as the Earth's at sea level. The Orbiter circulates its air through filters to remove carbon dioxide and other impurities.

This shot — taken from a satellite launched from *Challenger's* payload bay — clearly shows the Orbiter's spindly robotic arm, its silver - coated radiators, and the twin viewing ports set into the cabin roof. The apertures in the nose section are RCS (Reaction Control System) nozzles, which are used to make manouvers whilst in orbit.

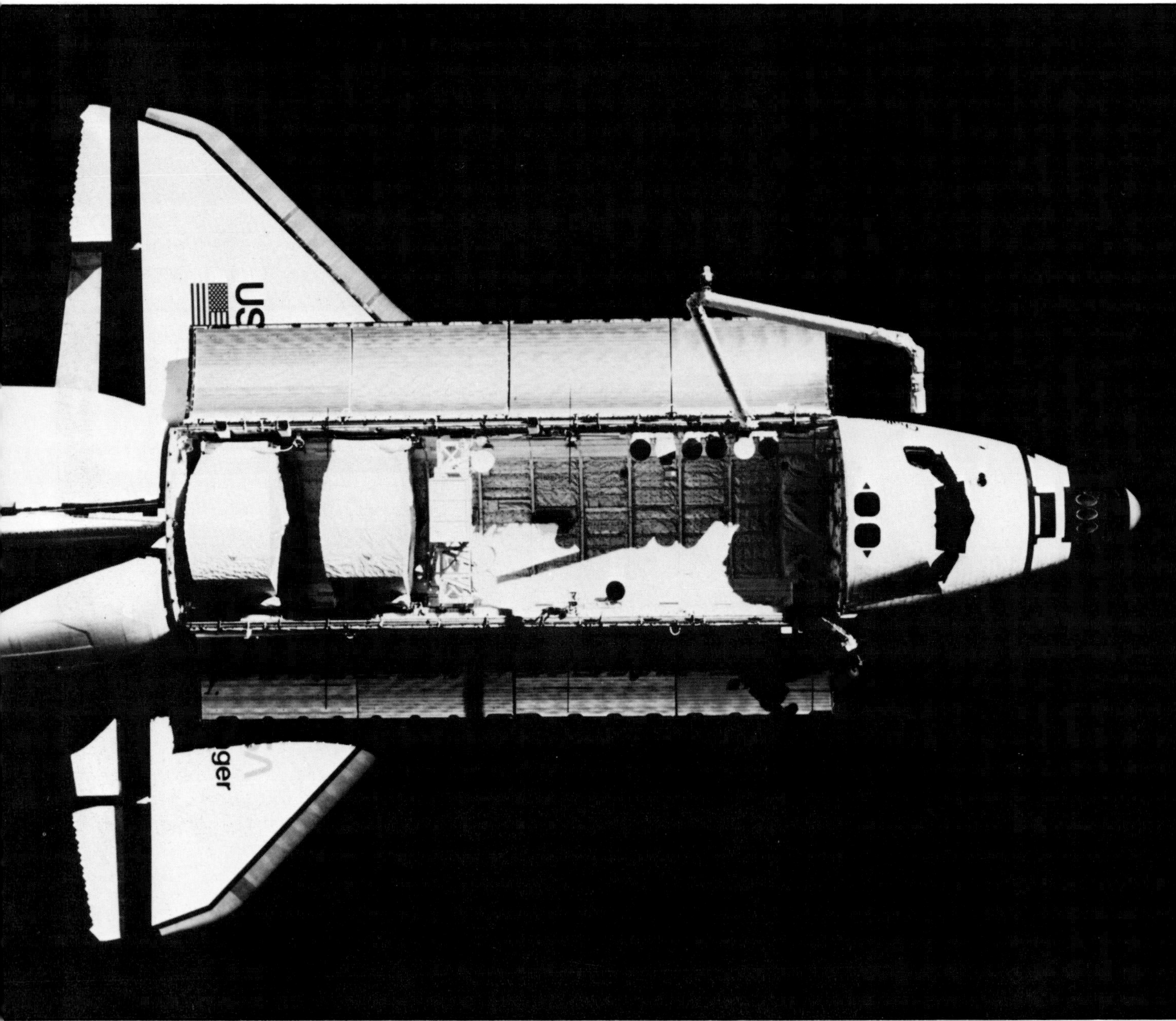

Excess moisture is also removed, keeping humidity at comfortable levels. Temperature in the Orbiter's cabin can be regulated between 16 and 32 degrees C (61 and 90 degrees F).

Living in the Orbiter requires only ordinary clothing; what NASA calls a 'shirt-sleeves' environment. Astronauts can move about, work and relax unencumbered by bulky spacesuits.

Foods for consumption in space have gradually evolved over the years since the days of the Mercury programme in the early sixties to become very appetising. Early Mercury flights were so short there was no need for food anyway, but experiments were nevertheless conducted by their crews with a view to the longer-duration missions everyone knew would follow. Before the first orbital flight there was even a question mark over the ability of a human to swallow food in zero-gravity, but experiments with bite-size cubes, freeze-dried foods, and semi-liquids in aluminium tubes proved that getting the food down presented no problem. Subsequent studies have shown that foods can be eaten with ordinary spoons and forks as long as there are no sudden starts, stops or spinning. The surface tension of the food holds it together and on the utensil. As a result, dining in space is almost like dining on Earth. But none of the early astronauts described their food in glowing terms; both taste and texture left a lot to be desired. They also experienced difficulty in rehydrating dried foods, and wasted a fair amount of valuable time trying to recapture free-floating crumbs before they found their way into delicate instruments.

By the time the Gemini flights came around in 1965, the food situation had improved somewhat. Aluminium squeeze tubes were abandoned because the bulk of their weight comprised the container rather than the contents, and excess weight aboard spacecraft is of course carefully avoided. Crumbling was controlled by coating the bite-sized cubes of meat, fruit, bread and dessert with a thin coating of edible gelatine. Dehydratables were by this time contained in plastic packs with a one-way valve at one end and a tube at the other. By inserting the nozzle of a water-gun device into the valve and adding a squirt or two to the contents, then kneading the mixture into a puree that he could squeeze through the tube into his mouth, the astronaut could keep his hunger at bay.

Step by step, through the Apollo, Skylab and U.S./Soviet Apollo-Soyuz missions, the groundwork was laid for the enticing range of foods that are now available to Space Shuttle crews. The Orbiter has its own galley located in the dining area on the lower level of the cabin, where food is stored, prepared and served in a level of comfort that would have made the Mercury crews green with envy.

Although the Orbiter does not have refrigerators or food freezers as Skylab had, it does incorporate something new to spaceflight — a convection oven. It also has hot and cold running water, food serving trays and a personal hygiene station where crewmembers may wash their hands before a meal and clean their teeth after it. There is even a table and attached restraints which function as chairs in the Orbiter's zero-gravity environment, though the phenomena of weightlessness allows them to attach their food trays to the walls and eat off them there if they prefer! The Space Shuttle's menu is very comprehensive; seventy-five different kinds of food and twenty beverages. Menus provide about 3,000 calories daily, previous space missions having demonstrated that astronauts need at least as many calories in space as they do on Earth.

Culinary highlights include broccoli au gratin, asparagus, turkey tetrazzini, ice cream, beef almondine and shrimp cocktail, while beverages range from cocoa to tropical punch. The menu is repeated every six days.

Food is eaten directly from packages containing individual servings which are assembled into balanced meals, over-wrapped in pouches, and stowed in locker shelves before launch. Each day, a certain crewmember is assigned to the task of meal preparation. It takes one crewmember about twenty minutes to prepare a meal for a crew of seven. Once the meal packages have been removed from storage, rehydratable food items are made edible by means of a sharp needle that injects water into their plastic packages. When the contents have been rehydrated, the top is cut off with a pair of scissors and the contents are ready to be eaten. Rehydratable foods outnumber the other types of food in the Orbiter's galley because they offer considerable savings in both weight and space. The water is in plentiful supply because it is a by-product of the Orbiter's fuel cells which generate electricity, so all in all it's an extremely elegant arrangement. Despite the suitability of the 'rehydratables', as NASA call them, not all types of food lend themselves to this

The Space Shuttle's commode and urinal — note foot restraints and 'seatbelt' for zero-gravity use! Unlike previous American spaceflight programmes women astronauts are participating, so the waste management system must be suitable for them too.

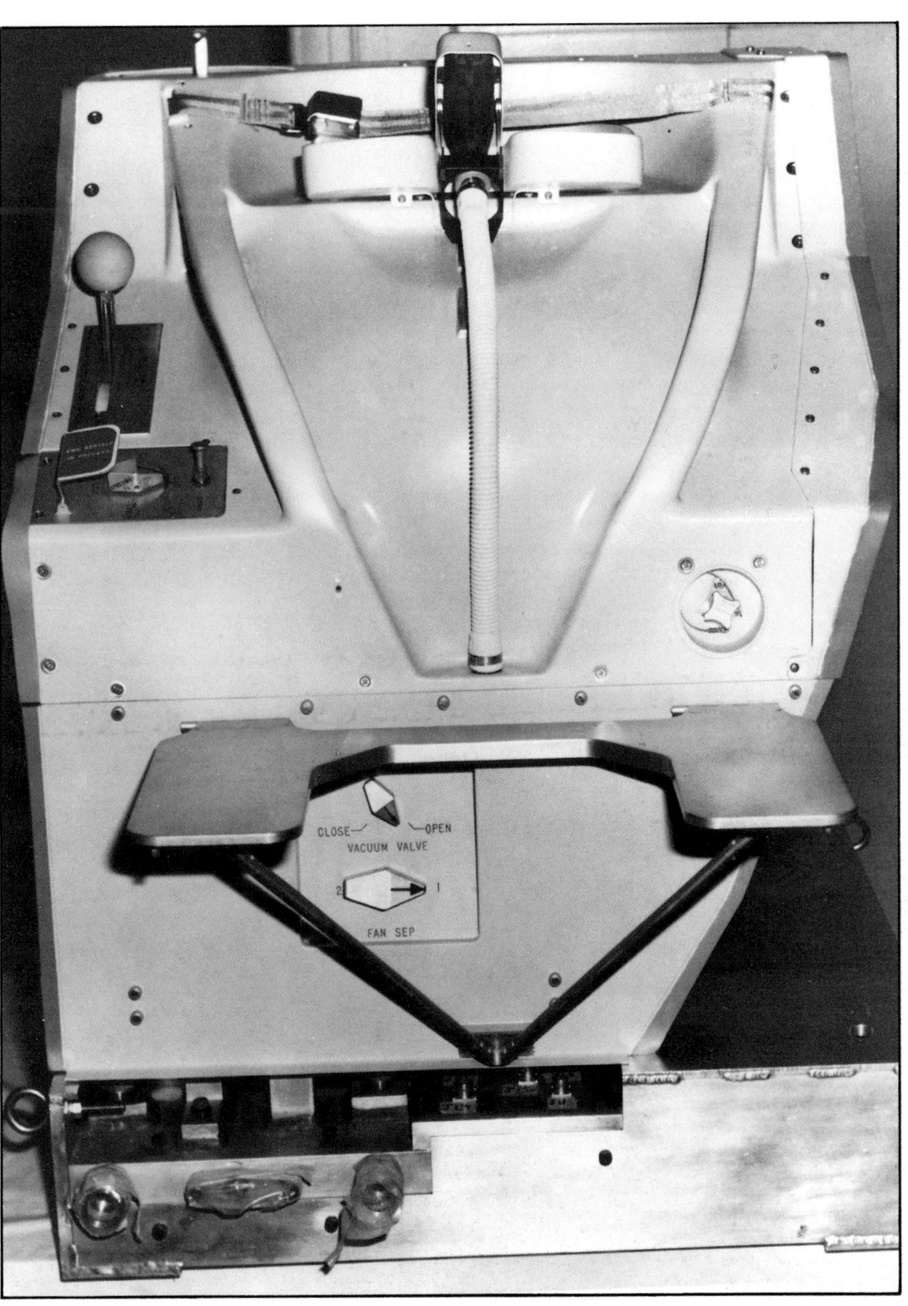

type of preservation. For this reason, a fairly large number of items fall into the 'thermostabilised' group. These foods have been precooked, with their temperature reduced to 4.4 degrees C (40 degrees F) prior to launch to increase their shelf life. Examples of thermostabilised foods carried aboard the Shuttle include ham, stewed tomatoes and lemon pudding. A small number of foods are best kept in their natural form, such as nuts, bread and cookies, while pears are freeze-dried prior to a mission.

In previous programmes, the menu varied according to personal preference, whereas Shuttle crewmembers eat the same standard meal. This partly reflects the 'workaday' operational nature of the Shuttle programme, as opposed to the one-off experimental nature of, say, the Apollo programme, though there are some pantry-stocked foods aboard the Orbiter which the crewmembers may select from if they do not desire a particular item on the menu. After a meal, the empty food containers are stowed in a trash compartment and the eating utensils, which are exactly the same as those used on Earth, are cleaned with wet-wipes for later re-use. The difference between Orbiter wet-wipes and those used on Earth is that the Orbiter's contain a strong disinfectant.

Sanitation is even more important within the confines of the Orbiter than it is on Earth. Space studies have shown that the population of some microbes can increase extraordinarily in a confined weightless area such as a spacecraft cabin. This could potentially spread illness to everyone on board. As a result, not only the eating compartments but also the dining area, toilet and sleeping areas are regularly cleaned. Since there are no washing machines in space, trousers (changed weekly), socks, shirts and underwear (changed every two days) are sealed in airtight plastic bags after being worn. Garbage and trash are also sealed in plastic bags.

Astronaut Rhea Seddon washing her hands in zero-gravity conditions aboard one of NASA's KC-135 aircraft, which is flying a parabolic curve to produce brief periods of weightlessness. The Orbiter's equivalent of a wash basin employs an airflow to cause water to rush across the user's hands like water flowing from a conventional tap.

This picture shows Astronaut Rhea Seddon attaching prewrapped food packages to a serving tray by means of Velcro pads. Due to the lack of gravity, astronauts may attach their food trays to the Orbiter cabin's walls and eat off them there if they prefer! The Space Shuttle's menu is very comprehensive; seventy-five different kinds of food and twenty beverages. Menus provide about 3,000 calories daily.

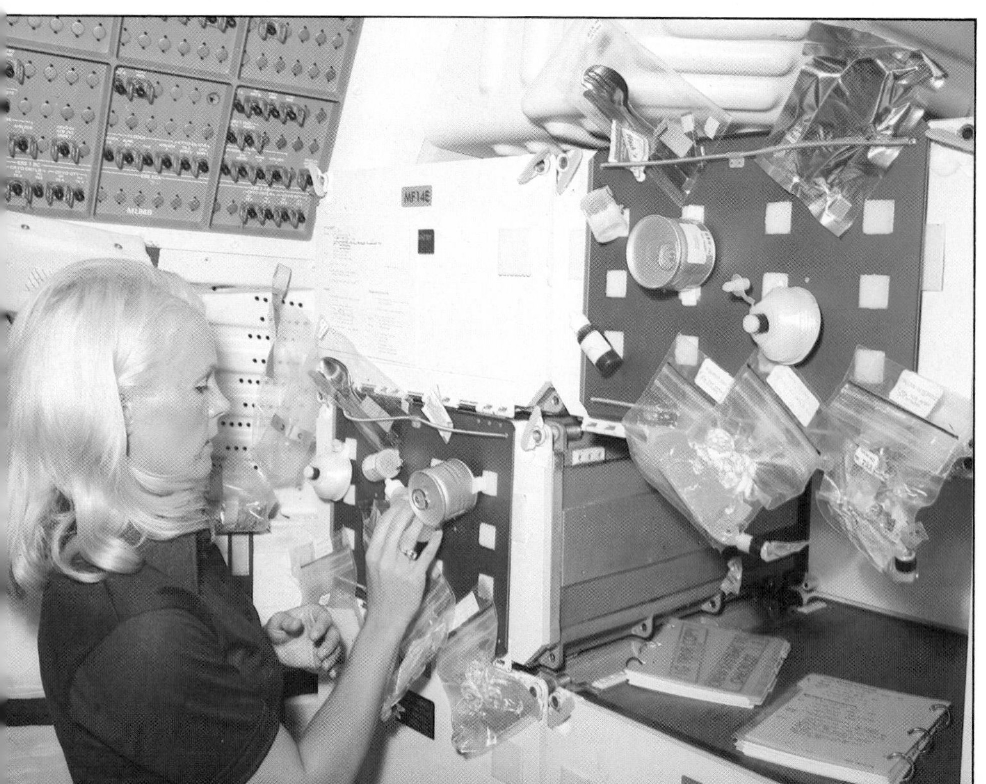

Orbiter travellers have facilities and supplies available for sponge baths while in space. They obtain water from a watergun, the temperature of which can be set at any comfortable level from 18 to 35 degrees C (65 to 95 degrees F). Because of weightlessness water droplets would float about in the cabin. This could be not only a nuisance, but also potentially hazardous to equipment and crew. To prevent this from happening, an airflow system directs waste water into the Orbiter's waste collection system, where it is sealed in watertight plastic bags. If whiskers cut off in shaving floated about weightlessly in a cabin, they could be a nuisance and even foul up equipment. An astronaut aboard the Shuttle avoids this problem by using conventional shaving cream and a safety razor, then cleaning off his face with a diposable towel. Also available is a windup shaver that works like an electric razor. It contains a vacuum device to prevent cut whiskers from escaping.

To wash their hands under zero-gravity conditions, astronauts use a device in

Astronaut Seddon with a typical Shuttle meal. Food is eaten directly from packages containing individual servings which are assembled into balanced meals, overwrapped in pouches, and stowed in locker shelves before launch. Each day, a certain crewmember is assigned to the task of meal preparation.

which an airflow causes water to rush across their hands like water flowing from a conventional tap, then into the Orbiter's waste water storage tanks.

Not unusually perhaps, one of the questions most frequently posed by members of the public when talking with astronauts at lectures and the like, concerns the arrangements for coping with the call of nature. Back in 1961 when Alan Shepard made his triumphant assault on the edge of space in Mercury-Redstone 7, no-one had anticipated that he might have a full bladder to contend with. After all, the flight was expected to last no more than fifteen minutes. But pre-launch delays meant that poor Shepard, lying on his back in the cramped confines of his space capsule, had to wait four long hours before finally getting airborne. He sheepishly explained his pressing need to doctors on the launch pad over a private communications line and, after a short delay whilst hurried consultations were made, received the message "Do it in your suit". And so it was that the first American in space made that momentous journey soaked in a pool of his own urine . . .

Thereafter, Mercury astronauts were fitted with a condom held in place by a pair of womens tights. The missions were comparatively short so there was no need to provide bags for defecation. Crewmen ate a low-residue diet for three days prior to launch to help them make this practical solution.

There has been a great deal of progress in this area since those somewhat primitive pioneering days. The waste management system employed on the Space Shuttle has been designed to be as routine and Earthlike as possible. Unlike previous American space programmes, women are flying as astronauts on a regular basis, so the new system must be suitable for them too. Before the final design was arrived at, twelve female and seven male subjects participated in a test of the Shuttle's waste management system under both normal gravity and simulated zero-gravity conditions. Zero-gravity was achieved aboard a NASA KC-135 aircraft flying a series of long parabolic curves. This study resulted in a design which comprises a pipe fitted with a unisex cup for urination, plus a commode unit. Both of them work on the principle of employing a fast-moving current of air in place of gravity, drawing wastes — including tissue wipes — downwards into a storage area where they are vacuum-dried. Stored biowastes are not dumped overboard as some people have suggested, but are in fact removed after landing. Some of the waste may be used for post-flight laboratory analyses. In the past, such analyses have told doctors which minerals are lost excessively in space, and have helped to increase their understanding of bodily functions. In deference to zero-gravity, hand holds, foot restraints, and even a waist restraint similar to a car seat belt, help the user to maintain a good seal with the commode seat, but otherwise the procedure is very much the same as that employed on Earth.

Although a malfunction in the commode unit forced astronauts Lousma and Fullerton to resort to the tried and tested expedient of plastic bags on the third Space Shuttle flight, there is no doubt that — sanitation-wise at least — the modern astronaut has never had it so good.

Just as on Earth, recreation and sleep are vital to good health in space. Cards and other games, books, writing materials, and tape recorders and tapes to chronicle personal observations and impressions are available to Shuttle astronauts to help them relax between work and sleep periods.

One astronaut has compared sleeping in weightlessness to lying on a three-dimensional waterbed. Two basic systems for sleeping are available on the Orbiter; sleeping bags, as made available on the early test flights, and 'rigid sleep stations', available on later operational flights. Once the Orbiter became fully operational, either configuration could be flown. The cocoon-like sleeping bags are almost identical to those used on the Apollo and Skylab missions, except that they employ perforated Nomex material for better temperature control. Eye masks and earplugs are provided for all sleeping bag occupants to block out light and background noise, such as radio static and automatic OMS 'burns'. In zero-gravity, with no 'right way up', the Shuttle astronauts are oriented in the bags as if they were sleeping standing up!

Four rigid sleep stations are provided in the Orbiter's cabin; three stacked horizontally and one installed in a vertical position (relative to the Orbiter's orientation on a runway back on Earth). Each station is provided with a sleep pallet, a sound suppression blanket, sheets with zero-gravity restraints, an overhead light, a communications station, a fan, and even a pillow.

In space, exercise is just as important as sleep. A scientifically planned exercise programme is provided, largely as a countermeasure for the atrophy of muscles in a weightless environment. Many of the problems of going into space have been resolved, but the physiological effects of weightlessness are still not completely understood. Among them are leaching of minerals from bones, reduction in the rate of bone formation, atrophy of muscles when not properly exercised, and motion sickness. Motion sickness can afflict the most unlikely people. Skylab astronaut William Pogue, for example, was a member of the USAF's *Thunderbirds* aerobatic team and was virtually immune to motion sickness on Earth. But within hours of

Rescue in space

There is a very real possibility that a rescue mission will be necessary at some stage in the Shuttle's period of service. The 34-inch diameter personal rescue enclosure (PRE) illustrated in the three pictures to the right would be used in the event of an emergency in orbit, if non-spacesuited astronauts had to be moved from one spacecraft to another. The rescue ball contains its own short-term, simplified life support and communications systems. It was conceived and fabricated by members of Johnson Space Center's Crew Systems Division and has three layers; Urethane, Kevlar and, on the outside, a thermal protective cover to shield the occupant from the Sun's searing rays. There is a small viewing port in the PRE to relieve the claustrophobic effects of prolonged use.

entering space, he was as sick as a dog.

Weightlessness itself causes immediate physiological changes. Blood rises up into the chest and head, making the face puffy and causing nasal stuffiness, which often leads to a headache. As these effects take place, the legs become skinny. All the deleterious effects astronauts have experienced from zero-gravity have so far reversed after their return to normal gravity on Earth. In addition, some of the effects have been countered in space by exercise and food supplements. However, even vigorous exercise in space does not appear to stop bone loss or the decrease in the rate of bone formation. As a result, NASA is engaged in an intense and sustained effort aimed at understanding the cause underlying these changes, and then developing ways to prevent them. The increased information about bodily functions derived from this effort will pave the way for prolonged missions in space and ultimately contribute to our understanding of the physiology of all living things on Earth.

When an astronaut has to work outside the Orbiter — to 'go EVA' (extravehicular activity) — he or she must of course don a protective spacesuit. In the past, spacesuits were tailor-made for each astronaut; a time-consuming and costly process. The Shuttle spacesuit is manufactured in small, medium and large sizes, and can be worn by both men and women. The suit comes with an upper and lower torso, equivalent to a shirt and trousers. The two pieces snap together with sealing rings. A life-support system is built into the upper torso, whereas previous pressure suits had separate life-support systems which had to be connected up by means of an umbilical.

Only two suits are provided on each Shuttle flight. They are assigned to the pilot and the ranking mission specialist (the astronaut responsible for managing Orbiter equipment). For emergency use, spherical personal rescue enclosures capable of providing life-support and communications facilities to one occupant are available for the other crewmembers. If an Orbiter becomes inoperable in space and cannot return to Earth, the non-suited crewmembers would enter their personal rescue enclosures and the pilot and a mission specialist would transfer them from the disabled vehicle to a rescue craft, which would be sent up from either the USA or the USSR, depending on the circumstances. The spacesuited astronauts can accomplish the transfer by carrying the enclosures, by attaching them to the remote manipulator arm of the rescue ship (normally used for deploying payloads) or by rigging a 'pulley and clothesline' device between both spacecraft and attaching the enclosures so they can be pulled from one spacecraft to the other.

Such a spectacular rescue bid would doubtless grab headlines the world over. It is probably reasonable to speculate that just such an operation will be necessary at some stage in the Shuttle's projected 20-year lifespan. The prospect of the Soviets becoming involved in a future space rescue mission is a fascinating one. Already, the compatible docking system developed for the joint U.S./Soviet Apollo-Soyuz Test Project in 1975 is used on the Space Shuttle and is earmarked for future Soviet manned spacecraft. Thus American astronauts will be able to aid Soviet astronauts, too. If other nations choose to employ this system in the future, additional international space rescue capabilities will be ready 'just in case'.

To increase the degree of manouverability an astronaut has once he has left the confines of the Orbiter, a one-man propulsive backpack called the MMU (Manned Manouvering Unit) has been developed. The MMU owes most of its systems to development stemming from the Skylab M-509 astronaut manouvering unit (AMU), which was tested in the forward compartment of the Skylab orbital workshop during the three missions in 1973-4. An MMU snaps onto the back of the astronaut's spacesuit and enables him to travel well away from the Orbiter, or to inspect areas of the vehicle which are otherwise inaccessible. It will really come into its own during future orbital construction projects, and would be essential in a space rescue operation.

Space Shuttle astronauts don't always need to 'go EVA' to perform tasks outside the Orbiter. The Remote Manipulator System is a robotic arm about 50 feet (15 metres) long by 15 inches (0.4 metres) in diameter firmly anchored to the port side of the payload bay, which enables the astronauts to pick satellites out of the cargo hold and position them accurately in space. It is also able to grapple satellites already in orbit and place them in the cargo bay for return to Earth.

Developed by the National Research Council of Canada, with Spar Aerospace of Toronto as prime contractor, the Remote Manipulator System — or Canadarm as it's sometimes known — was funded to the tune of $100 million as Canada's donation to the U.S. space programme. Subsequent Canadarms are being sold to NASA for about $25 million apiece. A remote manipulator arm has joints at the shoulder, elbow and wrist, each operated by six small electric motors. At the 'working end' of the arm is the so-called end-effector; a robotic hand. By looking through the aft flightdeck windows, the astronauts have an unrestricted line of sight to most areas of the payload bay and operate the Canadarm by means of a joystick device. Cameras are fitted to various parts of the arm to provide closed-circuit television coverage of delicate operations. The first flight test

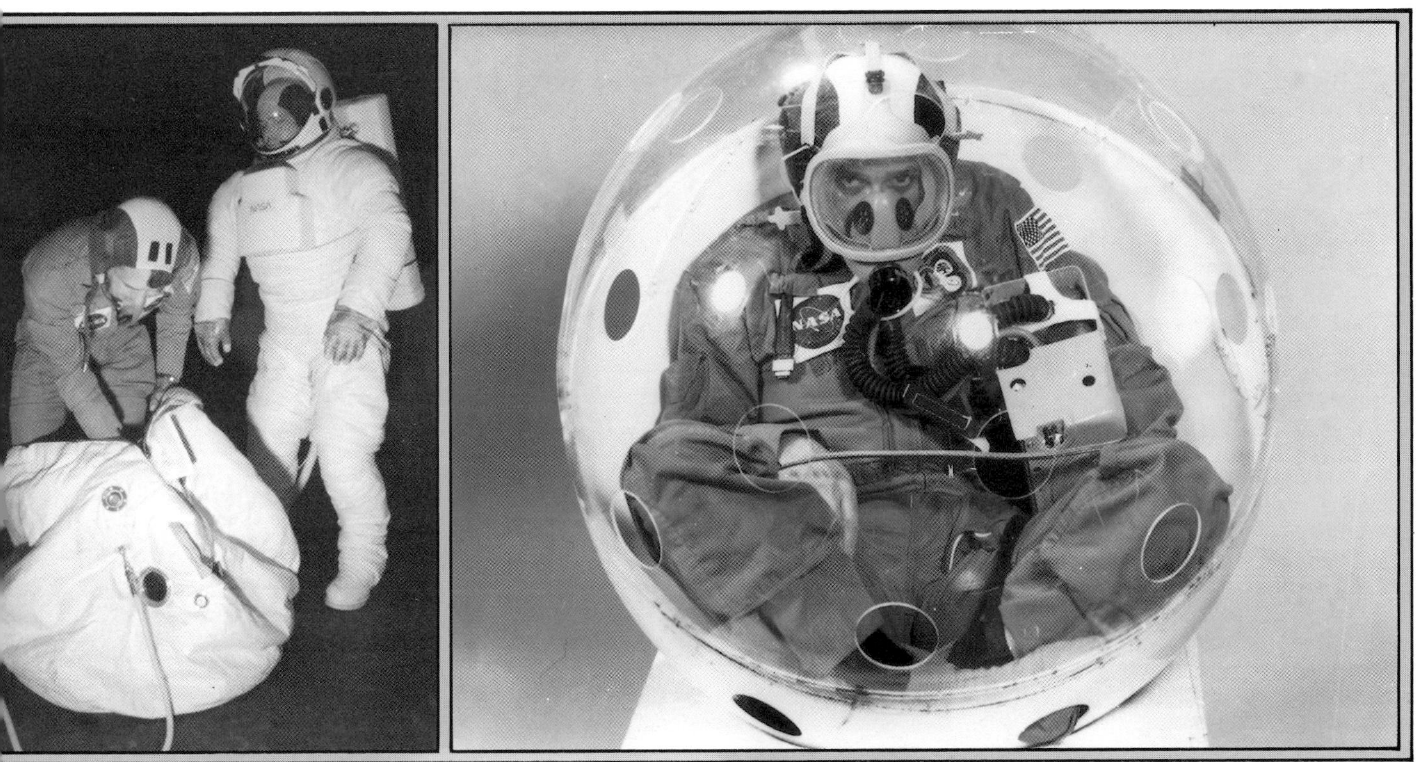

(Below) Astronaut Story Musgrave (a mission specialist on STS-6) goes EVA with a bag of latch tools. There is still a great deal to be learned about performing tasks in an airless environment, but steady progress is being made.

took place on STS-2. Operational duties in the future will vary from launching the huge Space Telescope to building gigantic solar power structures in orbit.

The Shuttle Orbiter carries a bewildering variety of payloads into orbit; military and civilian weather, navigation and communications satellites, experiments packages, space telescopes and telemetry equipment, to name but a few. The sheer diversity of experiments packages carried in the Orbiter's payload bay thus far prevent detailed description here, but two typically interesting studies were the SFXP and the MFE.

The SFXP (Solar Flare X-ray Polarimeter) was used for investigations into the nature of X-ray emission from solar flares. Such emission can be produced by beams of high-energy electrons and ionized gas raised to temperatures as high as 100 million degrees K. The SFXP was designed to discriminate between these two possibilities. Up until recently, the payload weight restrictions imposed by existing satellites have prevented any more than one analyser from being flown to investigate this phenomenon. No less than three were carried aboard the Shuttle Orbiter, for the most thorough sampling ever made of the characteristics of X-rays emitted during a solar flare.

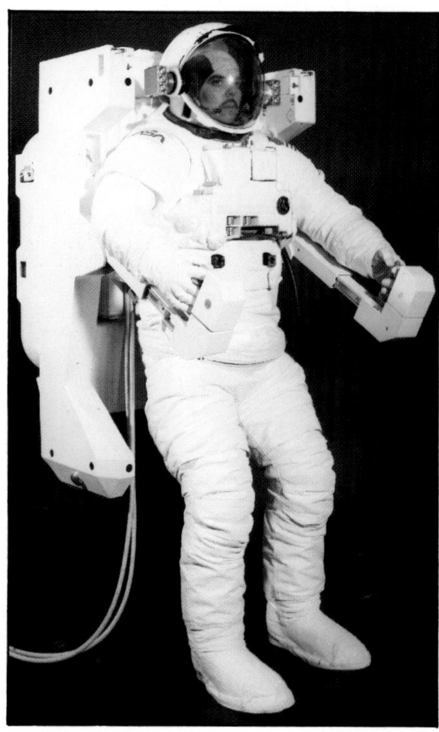

(Above, left and right) The MMU, or Manned Manouvering Unit is a one-man, self-contained backpack which enables Shuttle astronauts to repair Orbiting satellites, or rescue trapped crewmembers in the event of a Shuttle Orbiter being disabled. It will really come into its own during future orbital construction projects.

(Below) Thanks to an ingenious tether/slide-wire combination, STS-6 astronaut Donald Peterson (right) is able to translate along the port side handrails in the Orbiter's cargo bay. Meanwhile, Story Musgrave familiarises himself with working conditions in zero-G.

The MFE (Microabrasion Foil Experiment) was an equally fruitful Shuttle-borne project. Micrometeorites are encountered by all spacecraft in orbit around the Earth, and measurements of the numbers, chemistry and density of these tiny particles can yield basic information about the history of our solar system. Because comets and asteroids are formed in different regions of the solar system, it was anticipated that particulate materials from each source should differ. Comets were apparently formed at large distances from the Sun by the aggregation of ice and small dust grains. Asteroids, on the other hand, are believed to have been formed by the aggregation of stoney and metallic dust grains which condensed 4.6 billion years ago from the Mars-Jupiter region of the solar nebula.

Up until recently, our understanding of early solar system processes was based primarily on laboratory analyses of relatively large meteorites recovered on Earth. These analyses will always be deficient in studies of low-density cometary material, since meteorites made of such material hardly ever survive passage through the Earth's atmosphere. *In situ* collection techniques such as those employed on the MFE, however, provided a much-needed complimentary approach to extraterrestrial material studies.

The MFE was constructed of very thin aluminium foil sheets bonded to a gold-coated brass support mesh which was in turn bonded to a plastic (Kapton) substrate. These sheets were exposed to space during flight and examined after landing. Very light particles could not penetrate the foil, but left their mark in the form of an impact crater on the foil surface. Somewhat heavier particles could penetrate and be captured on the plastic backing layer. Heavier particles were not fragmented, surviving almost intact, whereas 'fluffy' or icy particles would readily fragment to form a number of impact craters in the Kapton backing sheet. An analysis of the fragmentation profiles on the Kapton sheet back on Earth therefore yielded accurate information on the particle densities. As for those micrometeorites which at least partially survived the impact, valuable chemical analyses were carried out on them.

(Above) Space Shuttle astronauts can perform some tasks outside the Orbiter without ever leaving the confines of its cabin, by using the Remote Manipulator System robotic arm. For example, it enables the astronauts to grapple satellites already in orbit and place them in the cargo bay for return to Earth.

(Above) This striking photograph was taken during the STS-6 mission by astronaut Karol Bobko from the aft window of the Earth-orbiting _Challenger's_ flightdeck. It shows mission specialist Story Musgrave in the Orbiter's payload bay testing the pull strength of the reel-in safety tether. The consequences of this system proving defective do not bear thinking about, but the astronauts have total confidence in their equipment.

The MFE was conceived by the University of Kent in Canterbury, England, which also analysed the returned foils subsequent to the test mission.

With the Space Shuttle being heralded as "the people's spacecraft", it was inevitable that a wider cross-section of the world community should be given access to its facilities. With a facility called 'Getaway Specials', NASA is allowing persons or groups able to meet certain requirements to fly their own small self-contained payloads on the Space Shuttle. This is the first time that such facilities have been made available to the general public. Since the offer was first announced in the autumn of 1976, literally hundreds of 'Getaway Special' reservations have been made by individuals and organisations

Technicians in the 'clean room' at NASA's Goddard Space Flight Center, Greenbelt, Maryland check out the installation of some scientific experiments aboard a Shuttle payload pallet. The U-shaped metal pallets are built by British Aerospace.

throughout the world. Industrialists, foreign governments, foundations, schools, colleges and universities, professional societies, service clubs and many others have staked their claim to a space in the Orbiter's payload bay.

There are no stringent requirements to qualify for a 'Getaway Special', but the payload does have to meet safety criteria. It must also have a scientific or technological objective, and be in good taste. Not "crassly commerical", to quote NASA's application form. For example, a manufacturer may want to test, on a proprietary basis, a certain type of metal-making process in space for later use in his own production. NASA state that this would be not only permissable, but welcome. However, under the "crassly commercial" prohibition, a person who wished to fly thousands of plastic discs for later resale as "objects that have flown in space" would be refused. As another example of a prohibited proposal, NASA say that a person who would want to have the ashes of a departed loved one placed in orbit via a 'Getaway Special' package would be denied space because it would be considered to be in poor taste . . .

'Getaway Special' spaces come in three standard sizes; 5 cubic feet, with a maximum weight of 200lbs. ($10,000), 2½ cubic feet and up to 100lbs. ($5,000), and 1½ cubic feet and up to 60lbs. ($3,000). NASA will provide advice on payload design, construction and testing and Shuttle crewmembers will turn on and off up to three payload switches, although generally there will be no opportunity for crew monitoring of 'Getaway Specials', or for any form of in-flight servicing.

As well as providing welcome access to space for a wide variety of people, the 'Getaway Special' scheme also has the benefit of earning worthwhile revenue from those spaces in the Orbiter's payload bay that are not occupied by the major payloads.

Not all Shuttle payloads are inanimate. In the past there have been monkeys and dogs in space, but the Shuttle programme has gone one stage further. The 'crew' of STS-3, for example, included three-dozen caterpillar moths, houseflies and honeybees. These insects owed their free trip to 18-year-old Todd Nelson of Rose Creek, Minnesota. Todd was the winner of a contest for U.S. secondary school students to suggest potential Shuttle experiments. He put forward the idea of finding out how various types of insects flew in zero-gravity. Unfortunately, most of them simply clung to the walls of their plastic cages, except when these were bumped into by the astronauts!

For many payloads, the Shuttle Orbiter's nominal 160-mile/maximum 600-mile orbit is just the first stage of a complex journey to a new home in space. A large proportion of the Shuttle's missions require the Orbiter to serve as a platform from which to launch other spacecraft — satellites — into their final, geostationary orbits 22,250 miles above the Equator. (A geostationary, or Clarke, orbit constantly maintains a spacecraft above one point on the Earth's surface).

To accomplish this goal, a device known as the Payload Assist Module (PAM) is employed. Two stages of propulsion are required once the satellite has been released from the Orbiter's payload bay. The PAM's first stage propels the satellite into a highly elliptical (or egg-shaped) transfer orbit, and is accomplished by means of a Thiokol-built Perigee Kick Motor. This is so named because the perigee — the point of the transfer orbit closest to the Earth — is the same as the Shuttle's orbit. Once in this transfer orbit, the satellite requires one more push to move it into a permanent circular orbit at its apogee — the point in the orbit most distant from the Earth. This final push is provided by another Thiokol-built unit in the PAM, an Apogee Kick Motor.

All satellites despatched into a higher orbit from the Orbiter's payload bay by means of the PAM are set spinning before their release to stabilise them. Powerful springs then push them well clear. The Orbiter turns its back on the satellite/propulsive stage combination to reduce the risk of damage as the first 'kick motor' ignites. The first Shuttle mission to actually release satellites in this way was STS-5, which deployed two of them on successive days. Up to five satellites can be deployed on a single mission.

In addition to these PAM-flown geostationary insertions, interplanetary spacecraft — exploratory probes — and larger geostationary payloads can be place in Earth orbit by the Shuttle, then blasted on their way by two large Inertial Upper Stage (IUS) motors, plus one small IUS.

Throughout the next decade, non-astronaut scientists, engineers, and technicians will be conducting in-orbit experiments that require close human supervision, thanks to Spacelab; a space-borne laboratory built by the European Space Agency (ESA). Set up on 31 May 1975, ESA groups in a single body the complete range of European space activities previously conducted by ESRO and ELDO in their respective fields of satellite development and launcher construction. Under the terms of its Convention, ESA's task is to provide for and promote — for exclusively peaceful purposes — co-operation among European states in space research and technology, with a view to their use for scientific purposes and for operational applications. Member states of ESA are Belgium, Denmark, France, West Germany, Ireland, Italy, the Netherlands, Spain, Sweden, Switzerland, and the United Kingdom. Austria, Canada and Norway participate in certain programmes; Austria is an Associate Member and is also participating in Spacelab, while Canada and Norway have observer status at the Council.

ESA has a total staff of about 1,425, located at headquarters in Paris and at its various technical establishments — at Noordwijk in the Netherlands, Darmstadt in West Germany and at Frascati near Rome. Several technical teams are located in national establishments for the conduct of certain space-related programmes.

ESA's Spacelab is carried aloft in the Orbiter's cavernous cargo hold for missions lasting up to nine days. A total of fifty of Spacelab missions are expected over the project's ten-year lifespan. The Shuttle/Spacelab combination saves money for experimenters because Spacelab's instrumentation and equipment is brought back to Earth and can be refurbished and reused, rather than abandoned in orbit. It also saves time; preparations for experiments aboard the Spacelab will be measured in months, rather than the years associated with automated satellites.

Spacelab offers pronounced benefits over the old Apollo and Skylab programmes. There is much more time available for experiments and better facilities than Apollo had to offer. Skylab did offer more room than Apollo, but the cost of getting crews to and from this orbiting space station was immense. Also, experiments could not be added or removed between missions and the experiment hardware

The positioning of the Spacelab orbital laboratory in the Orbiter's payload bay is shown to advantage in this Rockwell International artist's impression. An astronaut can be seen performing extravehicular activity (EVA) atop Spacelab's cylindrical habitable module.

returned to Earth. By comparison, Spacelab flights are shorter, but they are much more effective and are accomplished at a considerably lower cost. Scientists can now actually go into Earth orbit to carry out investigations for themselves, instead of relying on capable, but non-specialist, NASA astronauts as they did in the past.

Spacelab has facilities similar to laboratories on Earth, but adapted for operations in zero-gravity. It provides a shirt-sleeve environment similar to that in a commercial airliner, for both male and female experimenters. Initially, Spacelab will remain attached to the Orbiter whilst in space. Scientists eat and sleep in the Orbiter throughout the mission. On return to Earth, the laboratory is removed from the Orbiter and outfitted for its next assignment. Facilities are provided for as many as four laboratory specialists to conduct experiments in such fields as life sciences, plasma physics, pharmaceuticals, Earth observations, material sciences, solar physics, high-energy astrophysics and atmospheric physics.

Spacelab personnel are men and women from countries all around the world who are experts in their chosen field and are in reasonably good health. They require only minimal spaceflight training. The usual crew for a Shuttle/Spacelab mission consists of; two pilot astronauts, who operate the Shuttle vehicle and carry out the flight plan; one or two mission specialist astronauts, who operate the Orbiter/Spacelab interface systems and support payload operations; and up to four payload specialists — the individuals who actually operate the scientific equipment and conduct the experiments. A total of seven people may fly on any one mission.

Payload specialists are not 'career astronauts'. They may be selected to fly on more than one mission, but NASA has not created a permanent payload specialist corps.

A payload specialist's training includes visits to the investigators to observe and discuss all aspects of experiment development, and to practice performing the experiments using the actual equipment that will be flown in Spacelab later. The payload specialists also provide feedback to the investigators on how difficult the equipment is to operate. This helps refine equipment design and procedures and avoids operational problems later.

Spacelab's use is open to research institutes, scientific laboratories, industrial companies, government agencies and individuals all over the world. While many missions are government-sponsored, Spacelab is also intended to provide services to commercial customers. The new orbiting laboratory is capable of supporting almost any kind of experiment, including many which would be extremely difficult or even impossible on the ground. The absence of gravity in space allows scientists to conduct experiments that Earth's gravity interferes with. For example, many elements do not combine uniformly on Earth, because gravity causes the heavier molecules to settle to the bottom. In weightlessness they mix evenly. Crystals can be allowed to form in a purer state while floating in space, uncontaminated and unrestricted by the containers necessary to hold them under gravitational conditions. Lubricants, when weightless, spread over surfaces more evenly, and medical substances can be mixed more uniformly and produced with higher degrees of purity in the absence of gravity.

(Above) **An RCS (Reaction Control System) 'burn' being executed in orbit.**

Spacelab flies above most of the Earth's obscuring, filtering atmosphere, so any telescopes aboard will be able to see farther into space and with greater clarity, and its sensors can study gamma-rays and other forms of radiation from space; radiation which is essentially blocked by the atmosphere. Spacelab is also useful for Earth observations. This provides data for land-use studies, pinpoints pollution, spots crop and forest diseases, provides leads in the search for minerals, water and other natural resources, facilitates more accurate mapping and makes information available for many other services.

Spacelab consists of two main elements; a pressurised manned laboratory module and an external non-pressurised instrument platform, comprising up to five pallets. These British Aerospace-built pallets are for telescopes, antennas, and other large man-directed instruments requiring direct exposure to the space environment and a broad field of view. This modular design means that Spacelab configurations can be varied to meet the needs of many different types of missions;

(Below) **Whilst a Shuttle mission is underway, the Mission Control Center at Johnson Space Center, Houston, Texas watches closely.**

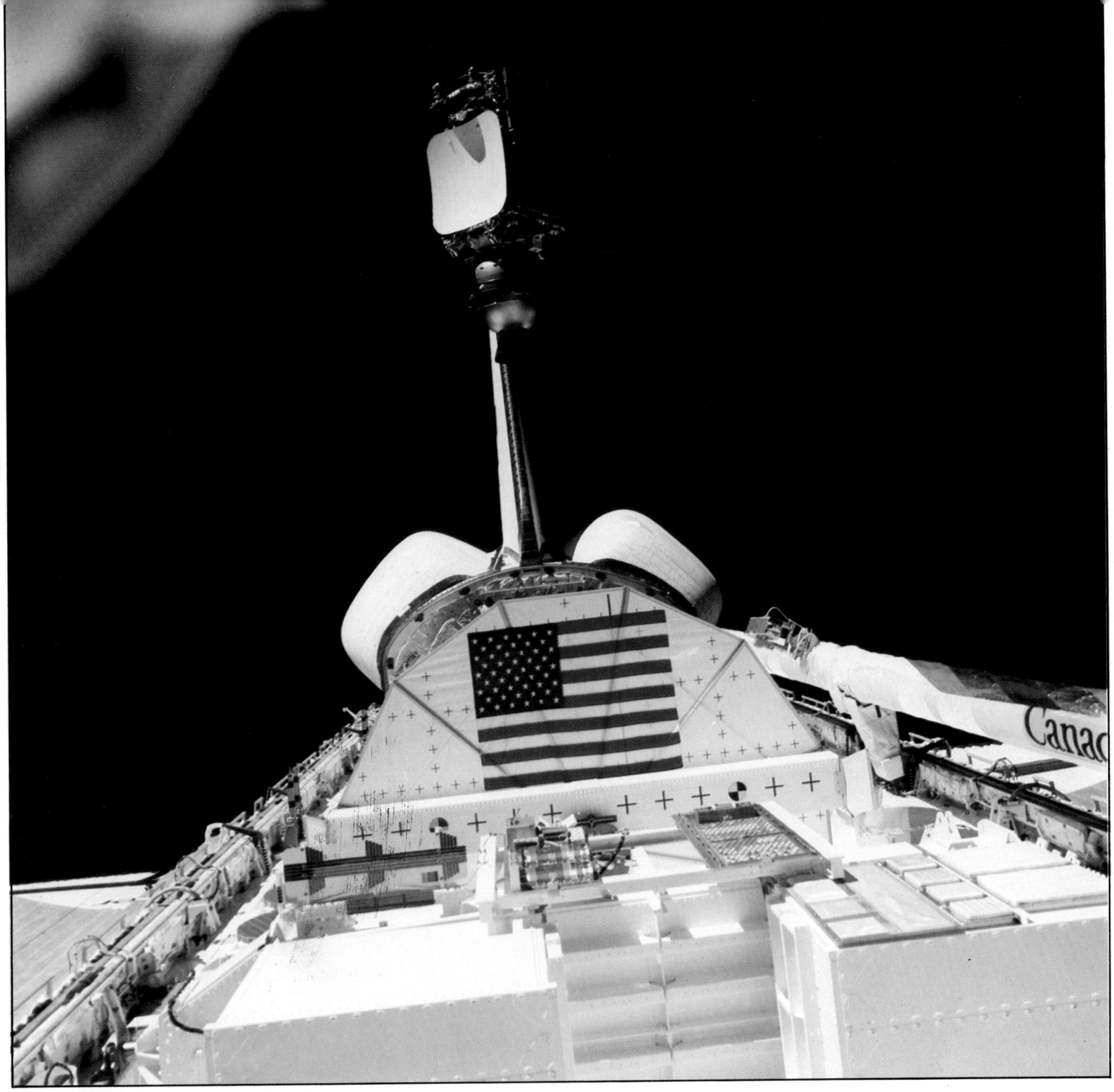

An Indian Government INSAT satellite can be seen departing above *Challenger's* tailfin.

module only, module plus pallet(s), and pallets only. There is also a self-contained instrumentation 'igloo' for use in conjunction with the pallets when the manned module is not flown, plus a tunnel one metre in diameter to link the laboratory with the Orbiter's cabin. The length of this tunnel varies from 4½ feet (1.4 metres) to almost 19 feet (5.7 metres), depending on the number of pallets carried behind the habitable module.

The prospects of Spacelab being improved are related to an increase in on-board electrical power, so as to provide greater flexibility of use. Such an improvement would open Spacelab to an even greater variety of missions, which could be extended from the present one-week maximum duration to a full month. In addition, it is planned to facilitate access of experimenters to Spacelab by means of an instrument pool — which could be used on several missions — and the establishment in Europe of a data-processing centre.

Finally, at a further stage Spacelab may become totally autonomous, and it may be possible for the pressurised module to be set into free orbital flight for very long-duration missions. This would thus represent the first element of a modular space station regularly visited by Space Shuttle Orbiters, which would serve as the basis for the construction and assembly in orbit of much larger space systems.

ESA prime contractor for Spacelab construction is ERNO at Bremen in West Germany, but individual firms in all ten member nations take part in the project. A total of about fifty companies funnel parts to Bremen for final assembly. NASA's Marshall Space Flight Center in Huntsville, Alabama, has responsibility for technical monitoring of the programme's design and development activity and for management of the first three Spacelab missions.

The first Spacelab mission took place in November 1983 aboard STS-9. A total of thirty-seven scientific and technological experiments — thirteen sponsored by NASA and twenty-four by ESA — were selected for this mission.

RETURNING TO EARTH

The Shuttle Orbiter *Columbia* has a vertical descent rate of 140mph — faster than a free-falling skydiver — as it approaches its runway.

Once a mission is completed, the flight crew aboard the Orbiter must ensure that all equipment is carefully stowed, and that the payload bay doors are locked firmly in their closed position in preparation for re-entry.

In the case of a landing at Kennedy Space Center, deorbit begins over the Indian Ocean just south of Sumatra — *half way around the world!* To initiate the deorbit sequence, the Orbiter is manouvered to travel tail-first. An Orbital Manouvering System (OMS) 'burn' in this position, retrorocket-style, slows the spacecraft down just enough to begin the descent back into the Earth's atmosphere. Once the deorbit 'burn' has been completed, the Orbiter is manouvered back into a nose-first orientation and the crew operates the aerodynamic control surfaces to ensure that the hydraulic system is functioning correctly. From 400,000 feet onwards, small OMS adjustments maintain the correct nose-up attitude as the Orbiter continues its earthward plunge, while on-board computers keep track of positioning, speed, altitude and orientation to ensure that the correct re-entry course is maintained.

Re-entry takes place about 3,500 miles from Kennedy Space Center (KSC) at a speed of approximately 17,500mph, twenty minutes after the completion of the tail-first OMS 'burn'. As the Orbiter's thermal protection system soaks up the blistering heat of friction with the atmosphere, the vehicle is surrounded by a cocoon of red-hot air. This forms a plasma layer around the speeding spacecraft, created by excited air particles which have ionised — forming an effective shield against radio waves. The result is a ten-minute communications blackout; a feature of all re-entries, and not just those involving the Shuttle.

On re-entry, the Orbiter's crew

experience a phenomenon that has become temporarily alien to them. Gravity. After their prolonged spell in weightlessness, the 2G (twice the body's normal weight on the ground) deceleration forces they encounter as the atmosphere beneath them becomes steadily denser, makes them feel very strange.

(Pictures this page) A unique scene greeted a large number of visitors to Edwards Air Force Base, California on 5 September 1983 when the STS-6 mission ended with the Shuttle's first - ever night landing. *Challenger's* wheels kissed the surface of Runway 22 with the minimum drama and showed the world that 'Touchdown' had replaced 'Splashdown' in the space programme vocabulary. Orbiters do not have landing lights, but powerful runway lights are employed.

Once the Orbiter has descended to an altitude of about 250,000 feet, the atmosphere has become sufficiently dense for some measure of aerodynamic control to be exercised. This is achieved by means of the large aft body flap beneath the main engines, and by the elevons, rudder and speed brake, though Reaction Control System (RCS) adjustments augment these aerodynamic inputs until the Orbiter has descended to 80,000 feet. From this point onwards the Orbiter is nothing more and nothing less than a big white glider, decelerating steadily but still travelling faster than any conventional aircraft ever could. The sonic boom which heralds its arrival can be heard on Miami Beach.

At an altitude of 20,000 feet the descent angle steepens to 22 degrees. The Orbiter's vertical descent rate at this stage is 140mph — faster than a free-falling skydiver! A preflare manouver is carried out at about 2,000 feet to reduce the glide slope to 1½ degrees, but the landing gear is not deployed until the last moment to reduce drag as much as possible. The wheels start to descend from their wells when the Orbiter is at 250 feet and do not actually lock into position until the vehicle is 100 feet off the ground, eleven seconds before touchdown. A final flare-out pulls the nose up seconds before the four Goodrich tyres on the main landing legs hit the runway. Touchdown speed varies from 213-236mph.

The early Shuttle test flights from KSC terminated at either the Rogers dry lakebed on Edwards Air Force Base in southern California or White Sands in New Mexico, where there is plenty of room for error. But it was always intended that future flights would end on more or less conventional runways at their original launch sites. To this end, the Orbiter runway at KSC is already operational, while Vandenberg Air Force Base in California is also being equipped with a long Orbiter-specification runway.

The runway at KSC's Shuttle Landing Facility is among the world's most impressive, both in terms of its sheer size and its superb quality. It is 4,572 metres (15,000 feet)in length — with a 1,000-foot overrun area at either end — and 91.4 metres (300 feet) wide. In general terms, that makes it roughly twice as long and twice as wide as a run-of-the-mill commercial runway, and appreciably larger than the larger-than-average landing facilities at such places as London-Heathrow, Rome-Fiumicino and Los Angeles International. KSC's runway was built between May 1974 and August 1976 on approximately 1,350 acres of land which, prior to its purchase by NASA in the early 'sixties, was primarily given over to agriculture. It has a northwest-southwest alignment, and is sixteen inches thick in the centre (diminishing to fifteen inches on the sides). Underlying the concrete paving on the surface is a six-inch thick base of soil cement, and as with a commercial runway, threshold and edge lights outline the field.

Incoming Orbiters must obviously be guided safely and with pin-point accuracy to the landing point. To achieve this, a

(Above) Accompanied by one of NASA's T-38 chase aircraft the Orbiter *Columbia* plummets towards its landing point and *(Below)* makes a perfect landing.

sophisticated Microwave Scanning Beam Landing System has been installed. Final guidance information is provided by ground-based components of this system located in small shelters just off the runway's west side. System components on the far side of the runway send signals which sweep fifteen degrees on each side of the landing path with directional and distance data. Simultaneously, signals from a companion shelter near the touchdown point sweep the landing path to provide elevation data up to thirty degrees. Equipment aboard the Orbiters receive this data and make the required adjustments to the glidepath. Needless to say, the glidepath approach must be perfect, as the Orbiters come in unpowered like giant gliders and cannot turn around to try another approach in the way that conventional aircraft would. Most approaches are flown on autopilot, with the crew monitoring the computers' performance and ready to take over control if the need arises.

Walking along KSC's Orbiter runway, the most striking feature is its heavily-grooved surface. These quarter-inch deep grooves have been cut in a crisscross

The T-38 chase aircraft employed by NASA for all Shuttle launches and landings are flown by the astronauts who have already, or will soon, fly in the spacecraft themselves.

Once it has rolled to a standstill on the runway, the Shuttle Orbiter is surrounded by a convoy of service vehicles which render it safe.

pattern along the entire length of the runway to form countless 1⅛-inch squares. (The KSC tourist's guidebook on sale at the nearby Visitor's Center will tell you there are some 8,450 miles of grooving here, and who am I to argue!) The grooves, together with the slope of the runway — twenty-four inches from centreline to edge — provide rapid drainoff of any water which may accumulate after a heavy Floridian cloudburst, thus preventing dangerous hydroplaning.

Although designed specifically for the Shuttle Orbiter, KSC's grooved runway can accommodate any commercial aircraft flying today or planned for the future. But it's not only aircraft which are attracted to its surface. Apparently, the local alligator population find it irresistable, too. Warmed through by the relentless Florida sun during the day, the runway stays warm well into the night — providing a perfect resting place for the alligators, which must control their temperatures very carefully in order to stay alive. Before an early-morning aircraft or Orbiter arrival, it is not unusual for a squad of KSC personnel to be despatched from the Industrial Area to shepherd the lazy 'gators to a safer refuge in the surrounding undergrowth!

When the Orbiter rolls to a standstill it is met by a recovery convoy whose primary functions are to provide immediate cooling and other vital services to the spacecraft, assisting crew egress, and towing the vehicle to the Orbiter Processing Facility, where it is deserviced before being prepared for its next journey into space.

As early as twenty-six hours prior to the Orbiter's spectacular arrival, the recovery convoy begins the vital preparations that will ensure a safe conclusion to the mission. These preparations begin with the warming up of coolant and purge equipment, followed by systematic testing of all ground servicing equipment and the convoy's own communications system. When the message finally comes through that the Orbiter's landing is expected in two hours, the convoy makes its way along KSC's taxiways to a standby parking area within sight of the runway. At landing minus five minutes a 'go status report' is conducted to ascertain final convoy readiness.

Initially, the Orbiter is approached with great caution. The convoy, positioned some 1,250 feet upwind of the now stationary spacecraft is led by a safety assessment team in a special self-contained van equipped to provide atmospheric protection for its occupants. Their first task is to detect potentially inflammable vapours in the vicinity of the Orbiter's main engines. The risk of an explosion or of personnel being overcome by lethal gases is a very real one, and NASA cannot afford to ignore it. A huge fire tender is poised for action, just in case. Two consecutive readings from three locations are made to check for concentrations of the Orbiter's hazardous fuels; hydrogen, monomethylhydrazine, hydrazine and ammonia. If calm wind conditions threaten to leave these toxic gases floating in the air, the NASA recovery crew simply create their own breeze to do the job for them! This is accomplished by

(Above) Re-entry. At a speed of 17,500mph, friction with the denser levels of Earth's atmosphere causes the Orbiter's undersides to glow red-hot and puts the thermal protection tiles to a severe test.

(Above) Rollout. The Orbiter's wheels do not actually lock into position until the vehicle is 100 feet off the ground, eleven seconds before touchdown. Touchdown speed varies from 213-236mph, depending on prevalent wind conditions.

(Below) When the Orbiter rolls to a standstill it is met by a recovery convoy which provides immediate cooling to the spacecraft's electrical systems, 'sniffs' for toxic gases, and safes explosive devices on board.

Just like a standard airliner, the Orbiter is departed by means of a long stairway.

means of a standard 14-foot diameter agricultural wind machine suitably modified for Shuttle landing operations and mounted on a mobile trailer unit. It can move a minimum of 200,000 cubic feet of air every minute.

Once the area surrounding the Orbiter has been declared safe enough for recovery operations to continue, the Purge and Coolant Umbilical Access Vehicles lumber into position at the rear of the Orbiter, the familiar Mack symbols on their blunt snouts somewhat incongruous in this Space Age setting. Before these are actually connected to the Orbiter by personnel in SCAPE (Self-Contained Atmospheric Protection Ensemble) suits, a hydrogen detection sample line is mated to yield information on the spacecraft's hydrogen concentration level. If this was found to exceed four percent, there would be an immediate power-down of the Orbiter and an emergency escape by the astronauts. Convoy personnel would then clear the area, thereby reducing the risk of heavy casualties in the event of an explosion. This would of course be an exceptional situation, and the hydrogen detection sample line procedure would normally be expected to indicate a level well below the four percent mark, thus allowing the Umbilical Access Vehicles to mate up with the Orbiter. The huge umbilical lines from these vehicles are connected to service points just beneath the spacecraft's bulbous OMS pods either side of the tailfin, and once *in situ* they perform their vital tasks.

The cooling operation is not unlike that carried out on a commercial airliner, and is achieved by pumping freon coolant along the umbilical lines and around the Orbiter's electronic equipment to prevent over-heating damage. The purge vehicle provides cool, humidified air-conditioning to the Orbiter's payload bay and other cavities to remove any residual explosive or toxic fumes and provide a safe, clean environment for later operations.

When further monitoring of vapour

When an Orbiter lands, these ground power and purging vehicles are attached to the aft end of the reusable spacecraft. The vehicles stay attached to the Orbiter and follow in convoy fashion as it is towed from the runway to the Orbiter Processing Facility (OPF).

levels around the Orbiter indicate that concentrations of toxic gases have reached a safe level, personnel in their baggy white SCAPE suits are replaced by somewhat more conventionally-dressed technicians. An environmentally 'clean' white room, identical to those used on the launch pads at KSC but mounted atop an airliner-type passenger stairway is now wheeled up to cover the crew access hatch. First man aboard the Orbiter is a physician, who carries out a preliminary medical examination on the astronauts, after which they are free to leave the confines of the spacecraft's cabin at last. They enjoy the fresh open-air breeze on their faces as they descend the long stairway, keeping a precautionary grip on the handrail as they find their feet again after the prolonged period of exposure to zero-gravity. Then they clamber into a special van, where they may remove their flight clothing on the way to the astronaut reception building.

Meanwhile, two other astronauts climb aboard the Orbiter and sit at the controls for the long, slow tow to the Orbiter Processing Facility, after a thirty-minute tyre cool-down period. During this wait, switch guards are installed in key positions on the Orbiter's instrument panels to ensure that potentially injurious operations (such as an OMS ignition) cannot be started accidentally. Later, these systems are drained of their propellants, and explosive devices are carefully removed. Another Shuttle mission is at an end.

When astronaut Bob Crippen peered out of his cockpit window as *Columbia* plummeted towards its very first landing, he cried out over the intercom, "What a way to come to California!" In spirit, the world replied as one; "What a way to go to space!"

Ground crewmen, garbed in SCAPE (Self-Contained Atmospheric Protection Ensemble) suits approach an Orbiter immediately after landing.

Post-landing activities include a shower for the ground crew. They are sprayed with water to wash off any hazardous substances they may have picked up on their suits.

THE ASTRONAUTS

(Above) This awesome photograph really needs no caption. Truly, one picture can be worth a thousand words. For the record, this is astronaut Robert Stewart using the jet-backpack MMU (Manned Manouvering Unit) during Shuttle mission 41-B/*Challenger* in February 1984. The Florida coastline is just visible beneath his feet.

(Far right) Astronaut Bruce McCandless stands on a device nicknamed the 'cherry picker' - officially, the Mobile Foot Restraint, or MFR - which can be mounted on the Shuttle's Canadian-built robotic arm. His feet are anchored into the MFR, allowing him to lean forward to perform various tasks.

(Below) Flash-fire damage was inflicted on this Shuttle spacesuit during a test. A technician received second-degree burns.

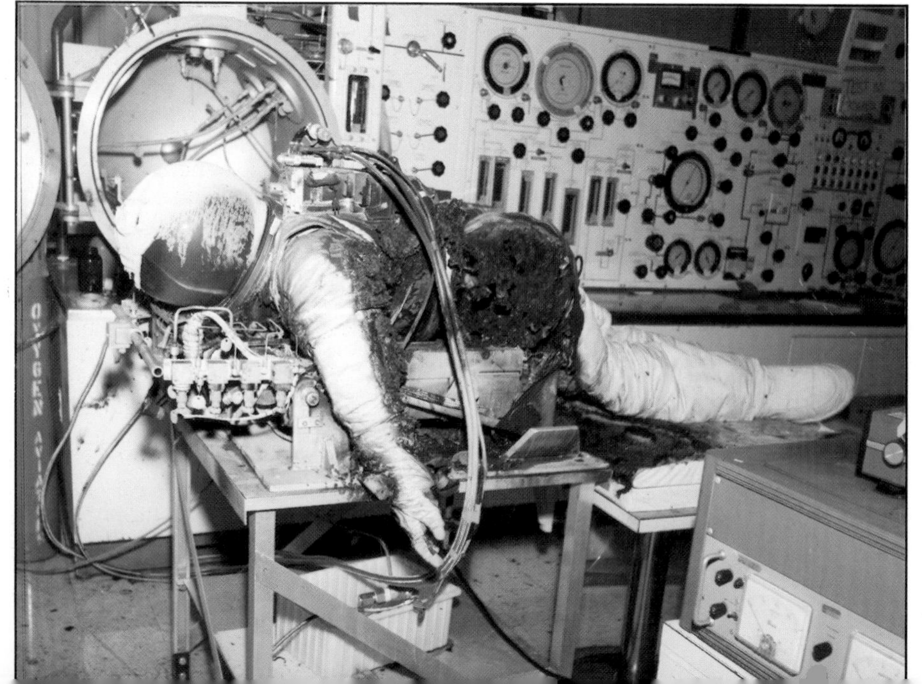

Spacemen of fiction — Jules Verne's travellers to the Moon, or the comic-strip heroes Flash Gordon and Buck Rogers — were familiar characters midway through the 20th. Century, but nobody could accurately describe a real astronaut. There were none to describe. Then, in 1959, NASA asked the United States military services to list their members who met specific requirements. The search was underway for pilots for America's exciting new manned spaceflight programme. In seeking its first space pilots, NASA emphasised jet aircraft flight experience and engineering training, and it tailored physical structure requirements to the small cabin space available in the Mercury capsule then being designed.

On 1st. August 1979, NASA announced plans to begin accepting applications by aspiring Space Shuttle astronauts on an annual basis. Depending on the space agency's needs for pilots and mission specialists, a selection is made from rosters of qualified applicants resulting from literally thousands of determined bids. Successful applicants are asked to report to the Johnson Space Center (JSC) — home of Mission Control — near Houston, Texas, for a one-year training and evaluation period as astronaut candidates, after which pilot and mission specialist astronauts are selected. Military applicants apply through the parent military organisation.

Minimum requirements for today's Space Shuttle astronauts include a bachelor's degree in engineering, physical science or mathematics, with advanced degrees desirable, and at least 1,000 hours of pilot-in-command flying time. High performance jet aircraft and flight test experience is highly desirable, and they have to pass a NASA Class I physical. The height requirement is between 64 and 76 inches. For mission specialists, the minimum requirements include a bachelor's degree in engineering, biological or physical science, or mathematics, with advanced degrees and/or higher levels of practical experience in their chosen fields desirable. The mission specialists are required to pass a NASA Class II physical, and they have to be between 60 and 76 inches in height.

Once selected to train as an astronaut, it's back to school studying basic science and technology courses, such as mathematics, meteorology, guidance and navigation, astronomy, physics and computers. The payload specialist receives the majority of his or her training from the payload developer, but receives approximately 150 hours of training at JSC. This training familiarises the payload specialist with the Shuttle vehicle and payload support equipment, crew operations, housekeeping,and emergency procedures related to his or her flight. To become accustomed to working in a pressurised spacesuit of the type worn when "going EVA" (extravehicular activity), astronauts spend many mission training sessions in one.

The training the pilots go through after becoming astronauts is constant and extensive. A good bit of preparation for the flights is done in simulators at JSC. One immensely complex simulator has prog-

rammed into it all possible sensations the astronauts experience during flight. Motion and vibration are as much a part of the simulation as the readings on the instrument panels. Even the gantry outside the port window of the vehicle moves during simulated launches. This simulator also duplicates jettisoning of the twin solid rocket boosters at twenty-seven nautical miles altitude. "The one thing that can't be simulated", says veteran Shuttle astronaut John Young, "is the G-forces through your chest. But the maximum we get is three. That's a piece of cake compared to the five or six Gs we got in the Apollo launches".

Whilst researching this book I paid a visit to JSC to take a closer look at some other astronaut training facilities. One of the highlights of my stay was being allowed to sit at the flight controls of a full-scale Orbiter nose section, which is normally positioned horizontally, but which can be tilted through ninety degrees to simulate conditions on the launch pad. It's a tight fit squeezing into one of the seats (ejection seats were fitted at the time), but once you get strapped in it's fairly comfortable. The Orbiter's cockpit features a truly bewildering array of instruments. These are situated not only in front of the crew on the main dashboard beneath the cockpit windscreen, but also down either side of the seats, all over the cabin ceiling, and even *behind* the seats! There are over 1,800 toggle switches, pushbuttons, rotary switches, circuit-breakers and other assorted controls on the Orbiter's flight deck. This is one hundred times the number found in an average car, and even three times that found in Apollo's Command and Service Module. New CRT (cathode ray tube) technology, which displays vital information onto a tv-type screen only when it is needed, has since been introduced into the Orbiter's instrumentation inventory, and this has reduced the profusion of dials and switches somewhat.

Adjacent to the tilting Orbiter nose section is the 'Orbiter one-G trainer'. This is a full-scale Orbiter fuselage, complete with flight deck and payload bay, which is used for flightcrew training in habitability, extravehicular activity, both normal and emergency ingress and egress, closed-circuit television operations, waste management, stowage, and routine housekeeping and maintenance. Like the tilting Orbiter nose section, the 'one-G trainer' may be only a mockup, but it is fitted out with authentic equipment and has a very realistic feel to it.

Nearby is the 'remote manipulator system task trainer'. This consists of an Orbiter aft crew compartment complete with aft cabin windows and a mechanically-operated arm. Astronauts practice payload grappling with the aid of full-scale helium-filled plastic models. Payload bay camera operations, berthing and visual operations can also be simulated on this trainer. Simulation of this kind is vital to the astronaut training effort at JSC, and there aren't many aspects of Shuttle operations not represented in some way or another by a full-scale working mockup. Among these is a Spacelab simulator.

The astronaut training facilities at JSC are simply too numerous to allow detailed description here, but of particular interest is the 'Orbiter neutral buoyancy trainer'. This is a huge cylindrical watertank, in which a full-scale crew cabin mid deck, a Shuttle airlock and payload bay area are immersed. Its purpose is to provide a zero-G environment for EVA training.

But not all training is conducted on the ground or in large tanks of water. Training to familiarise astronauts with the environment of space includes periods in weightlessness. Up to half a minute of zero-G is simulated when a Boeing KC-135 (derived from the commercial airliner but with all the passenger seats removed and converted into a 'padded cell' where the astronauts can float about freely), is flown on a parabolic trajectory to give an effect similar to that felt in a rapidly descending elevator. During this spell in zero gravity, the astronaut practices activities such as drinking, eating and using various types of equipment. Shuttle pilots often fly the KC-135s to build up their experience of handling a large aircraft.

The Shuttle Orbiter's landing approach has been described by one astronaut, with some justification, as "the ultimate toboggan ride". There can be no second chance when landing the world's largest glider, of course, so the astronauts must receive extensive training in the highly unconventional 'deadstick' approach procedure, to ensure that they get it absolutely right first time. To achieve this, two aircraft known as STAs, or Shuttle Training Aircraft, are incorporated into the astronauts' training programme. These machines are basically Grumman Gulfstream II executive jet aircraft, heavily modified to enable them to simulate the flight characteristics of the Orbiter during its landing approach. At the heart of an STA is

'Shuttlenauts'

NASA's Shuttle astronaut corps alters constantly, as new candidates enter and old campaigners depart for pastures new. These brief biographies provide an insight into what makes a 'Shuttlenaut'...

RHEA SEDDON

Name: Margaret Rhea Seddon (M.D.)
Birthplace and date: Murfreesboro, Tennessee. 8 November 1947.
Physical description: Blond hair, blue eyes; height 5 feet 3 inches; weight 110 pounds.
Education: Rhea graduated from Central High School in Murfreesboro, Tennessee in 1965. She went on to receive a bachelor of arts degree in Physiology from the University of California, Berkeley, in 1970, and a doctorate of Medicine from the University of Tennessee College of Medicine in 1973.
Marital status: Married to astronaut Robert 'Hoot' Gibson.
Children: Paul Seddon Gibson, 1982.
Recreational interests: Not specified.
Career background: Dr. Seddon was selected as an astronaut candidate by NASA in January 1978. In August 1979, she completed the mandatory one-year training and evaluation period, making her eligible for assignment as a Mission Specialist on future Shuttle flight crews. Seddon has served NASA in a wide variety of ways. She has helped develop Orbiter and payload software, has been involved in perfecting the Orbiter's avionics at the SAIL facility at Johnson Space Center, helped develop the Shuttle medical kit and checklist, and has served as a launch and landing rescue helicopter physician and support crew member for the STS-6/*Challenger* mission.

Rhea's first space flight was as a Mission Specialist on 51-D/*Discovery* in April 1985. She was responsible for deployment of the Leasat 3 satellite, RMS mechanical arm operations and the American Flight Echocardiograph Experiment. Her second space voyage was STS-40/*Columbia*, the Spacelab Life Sciences-1 (SLS-1) mission, scheduled for launch in the summer of 1991.

JEFF HOFFMAN

Name: Jeffrey A. Hoffman (Ph.D)
Birthplace and date: Brooklyn, New York. 2 November 1944.
Physical description: Brown hair, brown eyes; height 6 feet 2 inches; weight 160 pounds.
Education: Graduated from Scarsdale High School, Scarsdale, New York in 1962. Hoffman received a bachelor of arts degree in Astronomy (graduated Summa Cum Laude) from Amherst College in 1966, and a doctor of philosophy in Astrophysics from Harvard University in 1971.
Marital status: Married to the former Barbara Catherine Attridge of Greenwich, London, England.
Children: Samuel, 1975; Orin, 1979.
Recreational interests: Hoffman enjoys skiing, mountaineering, hiking, bicycling, swimming, sailing and music.
Career background: Dr. Hoffman's research interests lie in high-energy astrophysics, specifically cosmic gamma-ray and X-ray astronomy. His doctoral project at Harvard was the design, construction, testing and flight-operation of a balloon-borne, low-energy gamma-ray telescope. From 1972 to 1975, during three years of postdoctoral work at England's Leicester University, Hoffman worked on three rocket-borne payloads: two for the observation of lunar occultations of X-ray sources and one for an observation of the Crab Nebula with a solid-state detector and concentrating X-ray mirror. He designed and supervised the construction and testing of the lunar occultation payloads and designed test equipment for use in an X-ray beam facility which he used to measure the scattering and reflectivity properties of the concentrating mirror.

During his final year at Leicester, he was project scientist for the medium-energy X-ray experiment on the European Space Agency's Exosat satellite, and played a leading role in the proposal and design studies for this project.

Jeff Hoffman worked in the Center for Space Research at the Massachusetts Institute of Technology (MIT) from 1975 to 1978 as project scientist in charge of the A4 hard X-ray and gamma-ray experiment fitted to the HEAO-1 satellite, launched in August 1977. His involvement included pre-launch design of the data-analysis system, supervising its operation post-launch, and directing the MIT team undertaking scientific analysis of the flight data the experiment returned. He also was involved in analysis of X-ray data from the SAS-3 satellite being operated by MIT, performing research on the study of X-ray bursts. Dr. Hoffman has authored or co-authored more than 20 papers on this subject since bursts were first discovered in 1976.

Hoffman was selected as an astronaut candidate by NASA in January 1978. In August 1979, he completed the mandatory one-year training and evaluation period, making him eligible for assignment as a Mission Specialist on future Shuttle flight crews. During preparations for the first four 'developmental' Shuttle space missions, Dr. Hoffman worked in the Flight Simulation Laboratory at Downey, California, testing guidance, navigation and flight control systems. He has worked with the Shuttle Orbital Maneuvering System (OMS) and Reaction Control System (RCS), with Shuttle navigation, with crew training, and with the development of satellite deployment procedures. Hoffman served as a support crewmember for the STS-5/*Columbia* mission and as a CAPCOM (Capsule Communicator) for STS-8/*Challenger*.

Jeff Hoffman's first space mission was 51-D/*Discovery*, launched in April 1985. In an attempt to rescue the malfunctioning Leasat 3 comsat, Hoffman and David Griggs undertook the first unscheduled spacewalk of the Shuttle program. Hoffman's second space voyage was on the STS-35/*Columbia*/Astro-1 mission, dedicated to astronomy, which was launched in December 1990 after many months of technical problems.

STORY MUSGRAVE

Name: F. Story Musgrave (M.D.)
Birthplace and date: Boston, Massachusetts. 19 August 1935.
Physical description: Blond hair, blue eyes; height 5 feet 10 inches; weight 148 pounds.
Education: Musgrave graduated from St. Mark's School, Southborough, Massachusetts in 1953. He went on to receive a bachelor of science degree in Mathematics and Statistics from Syracuse University in 1958, a master of business administration degree in Operations Analysis and Computer Programing from the University of California at Los Angeles (UCLA) in 1959, a bachelor of arts degree in Chemistry from Marietta College in 1960, a doctorate in Medicine from Columbia University in 1964, and a master of science in Physiology and Biophysics from the University of Kentucky in 1966.
Marital status: Divorced from the former Carol Peterson of Beaverton, Oregon.
Children: Lorelei Lisa, 1961; Bradley Scott, 1962; Holly Kay 1963; Christopher Todd 1965; Jeffrey Paul, 1967; Lane Linwood 1987.
Recreational interests: Chess, flying, gardening, literary criticism, long-distance running, microcomputers, parachuting, photography, racquetball, scuba-diving, and soaring.
Career background: Following graduation from high school in 1953, Musgrave entered the U.S. Marine Corps and completed basic training at Parris Island, South Carolina. He completed training at the U.S. Naval Airman Preparatory School and the U.S. Naval Aviation Electrician and Instrument Technician School in Jacksonville, Florida, then served as an aviation electrician and instrument technician and as an aircraft crew chief whilst completing duty assignments in Korea, Japan, Hawaii, and aboard the aircraft-carrier *USS Wasp* in the Far East.

Musgrave has flown over 150 different types of civil and military aircraft, logging over 18,000 hours flying time, including over 6,500 hours in jet aircraft. He holds instructor, instrument instructor, glider instructor and airline transport ratings. An accomplished parachutist, he has made over 450 free-falls, including more than 100 experimental free-fall descents involved with the study of human aerodynamics. He holds an International Jumpmaster Class C license and was President and Jumpmaster of the Bluegrass Sport Parachuting Association in Lexington, Kentucky from 1964-67.

In 1958, Dr. Musgrave was employed as a mathematician and operations analyst by the Eastman Kodak company in Rochester, New York. He served a surgical internship at the University of Kentucky Medical Center in Lexington from 1964-65, and stayed on there until 1967, initially as a U.S. Air Force postdoctoral fellow specialising in aerospace medicine and physiology, then as a National Heart Institute post-doctoral fellow teaching and conducting research in cardiovascular and exercise physiology.

Dr. Musgrave has written 45 scientific papers in the areas of aerospace medicine and physiology, temperature-regulation, exercise physiology, and clinical surgery.

In August 1967, Story Musgrave was selected as scientist-astronaut by NASA, then undertook astronaut academic training and a year of military flight training. He worked on the design and development of the Skylab space station, was the backup science-pilot for the first manned Skylab mission (Skylab 2), and was a CAPCOM (Capsule Communicator) for the Skylab 3 and Skylab 4 missions. He was the Mission Specialist on the first and second Spacelab Mission Simulations, and participated in the design and development of all the initial Space Shuttle EVA equipment, including spacesuits, life-support systems, airlocks and the Manned Maneuvering Units.

From 1979 to '82, Musgrave was assigned as a test verification pilot in the Shuttle Avionics Integration Laboratory (SAIL) at Johnson Space Center. He has continued clinical and scientific training as a part-time surgeon at the Denver General Hospital, and as a part-time professor of physiology and biophysics at the University of Kentucky Medical Center.

Story Musgrave has undertaken three space missions. His first was STS-6, *Challenger's* maiden voyage, which launched in April 1983. The mission's primary objective was achieved when the first of NASA's TDRS tracking and data-relay satellites was deployed. Soon after, Musgrave and Don Peterson undertook the first spacewalk of the Shuttle program.

Musgrave's second space mission was 51-F/*Challenger*/Spacelab 2, the first use of European-developed Spacelab hardware in its pallet-only configuration. The eight-day flight began dramatically, with a two-engined final ascent, when the Abort-To-Orbit contingency was utilized.

In November 1989, Musgrave made his third space flight, on the classified STS-33/*Discovery* Department of Defense-sponsored mission.

a Sperry 1819B airborne digital computer coupled to the Gulfstream II's standard autopilot and instruments in such a way that every conceivable nuance of the spacecraft's extraordinary flight characteristics can be programmed into the bizjet's flight control system.

The cockpit of an STA has a split personality; it is effectively half Orbiter, half bizjet. In front of the left-hand seat, an Orbiter-style rotational hand controller replaces the usual control wheel, and other Orbiter displays are included in the cockpit for extra authenticity. In addition, the centre console houses the data entry panels that are used to 'dial-in' automatic simulation of Orbiter flight characteristics. These modifications are of course invisible to the casual observer, but there is a very obvious external indication that the STA is something more than just a Gulfstream II painted in NASA's neat red, white and blue livery. These are the direct lift controls and side-force generators protruding from the aircraft's underbelly, which pivot to produce lateral loads similar to those the pilot of an Orbiter encounters on his landing approach.

When an astonaut climbs aboard to begin a training session in the STA he occupies the left-hand seat, with the instructor to his right. After take-off, which is absolutely conventional, they climb to 35,000 feet, make last-minute checks to verify that all systems are functioning correctly, then pitch down into the simulated unpowered landing mode. The main landing gear is lowered, throttles are pulled back to idle, and the in-flight thrust reversers — their force controlled by the computer — are engaged to simulate the aerodynamic drag of the real Orbiter. All the time, the autopilot and the computer-controlled simulation system are working together to maintain the programmed speed of the Orbiter in its steep 24-degree descent, which brings the STA earthward at a dizzy 12,000 feet per minute. Meanwhile, the astronaut trainee practices the required hand controller inputs by following command bars on an altitude and directional indicator similar to that which was used in the NASA/Grumman Lunar Module which put the Apollo astronauts onto the surface of the Moon. He can either fly the approach himself, or monitor the progress of the automatic flight control system as it does the work for him.

This strange shot was taken aboard NASA's Boeing KC-135 jetplane while it was flying a curving 'parabolic trajectory' to create brief conditions of weightlessness for its occupants. During training, this familiarises astronauts with the sensations that becomes commonplace in orbit. Pictured here enjoying his first experience of weightlessness is astronaut Bill Fisher. His wife, Anna - another astronaut - is lifting him with one hand. Note the thick padding on the cabin walls and ceiling.

Once the approach profile and 'dead-stick' landing have been completed, the instructor takes over the controls and flies the STA back up to 35,000 feet to repeat the whole procedure. It's a demanding routine, but it pays handsome dividends in terms of increasing the astronauts' confidence. By using the STA, training for the actual event is done realistically, yet at far less expense than if an actual Orbiter was employed for the task.

NASA currently has a corps of about 100 astronauts. Its crew allocation procedure is a closely guarded secret, though it is known to involve seniority, and an attempt to match education and experience to the mission. After STS-4, NASA stopped naming backup crews for missions. There is now only a primary crew. I asked why this should be and was told, "It's due to the crunch in training time in the simulators versus the compactness of the schedule. There simply isn't time to train two crews for each mission in the short time available between Space Shuttle launches. Don't ask me what they do if somebody gets sick. Only Houston could weasel their way out of a question like that".

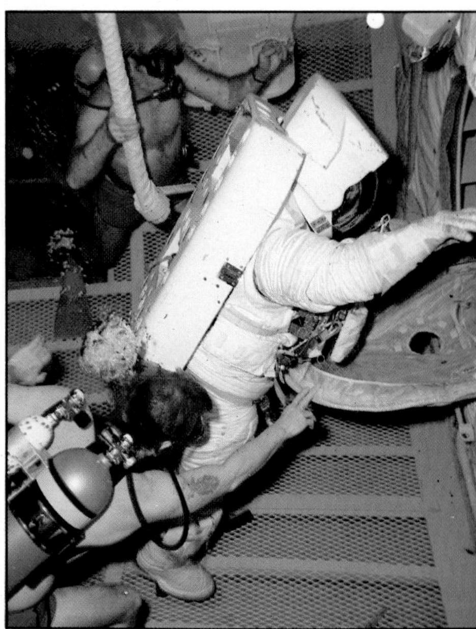

(Above) **Astronauts practice extravehicular activity (EVA) in large water tanks equipped with full-scale sections of the Space Shuttle Orbiter. In this watery environment, the astronauts are in a state of 'neutral-bouyancy' similar to that experienced in zero-gravity whilst in orbital flight. This way, a useful degree of familiarization is achieved at relatively low cost.**

(Above) The STA. Sophisticated onboard computers enable it to mimmic the handling and control responses of the Orbiter during the unpowered descent and 'deadstick' landing sequence. The cockpit has a split personality; Orbiter controls on the left-hand side (for the astronaut under training) and conventional Gulfstream 2 bizjet controls on the right (for the instructor-pilot).

(Left) An STA on a training flight, with a NASA T-38 chase-plane in close attendance.

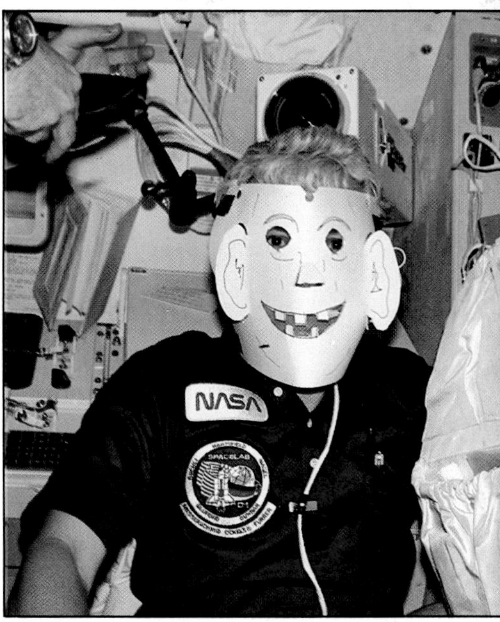

(Above) Fun and games are not exactly unheard-of on Shuttle flights! When the crew of mission 61-A/*Challenger* found themselves in orbit during Halloween, Hank Hartsfield got into the spirit of things with this home-made 'jack-o-lantern' mask.

SHUTTLE FLIGHT DECK

The Shuttle Orbiter's flightdeck has over eighteen hundred toggle switches, pushbuttons, rotary switches, circuit-breakers and other assorted controls. This Poster simply isn't large enough to accommodate the Shuttle's entire complement, but these are the main instrument panels facing the Commander and Pilot. The instruments not illustrated here are those installed in the aft flightdeck area and in the cabin mid-deck. The diagram immediately below provides a general layout of all the principal instrument panels.

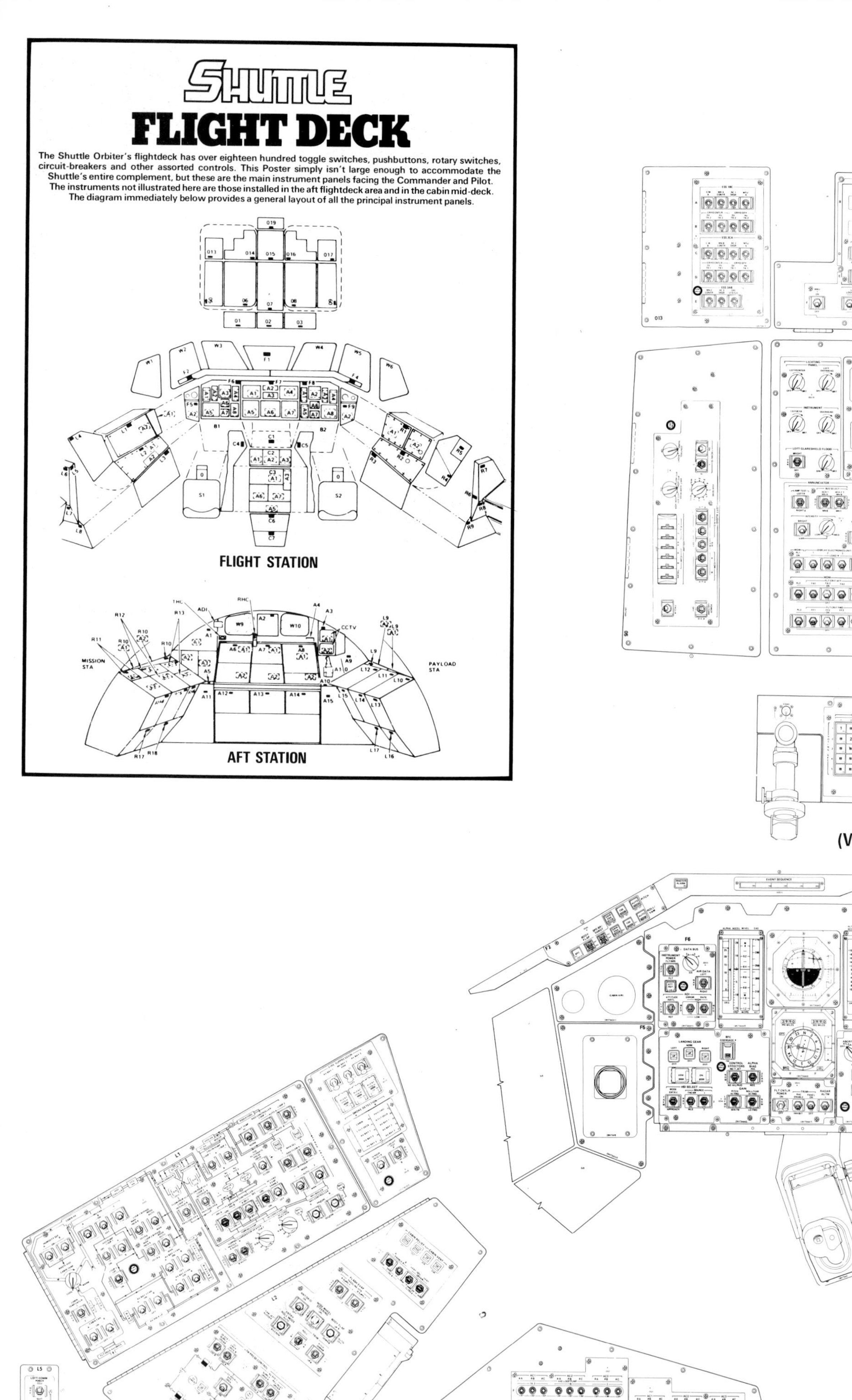

COCKPIT CEILING

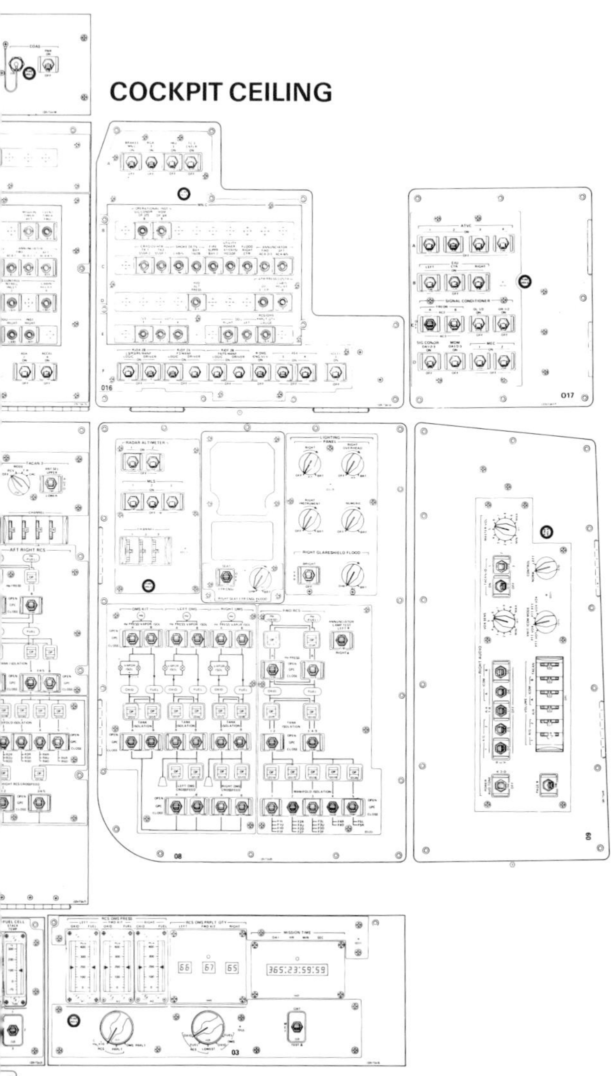

The Shuttle Orbiter's cockpit features a truly bewildering array of instruments. These are situated not only in front of the crew on the main dashboard beneath the cockpit windscreen, but also down either side of the seat, all over the flightdeck ceiling, and even *behind* the seats. The positioning of 'pilot' and 'co-pilot' on the flightdeck follows the time-honoured airliner practice, with the spacecraft's Commander in the left-hand seat and the Pilot to his immediate right. Instrumentation is sufficiently duplicated to permit the vehicle to be piloted from either side and facilitate a one-man emergency return to Earth.

(WINDSCREEN)

MAIN DASHBOARD

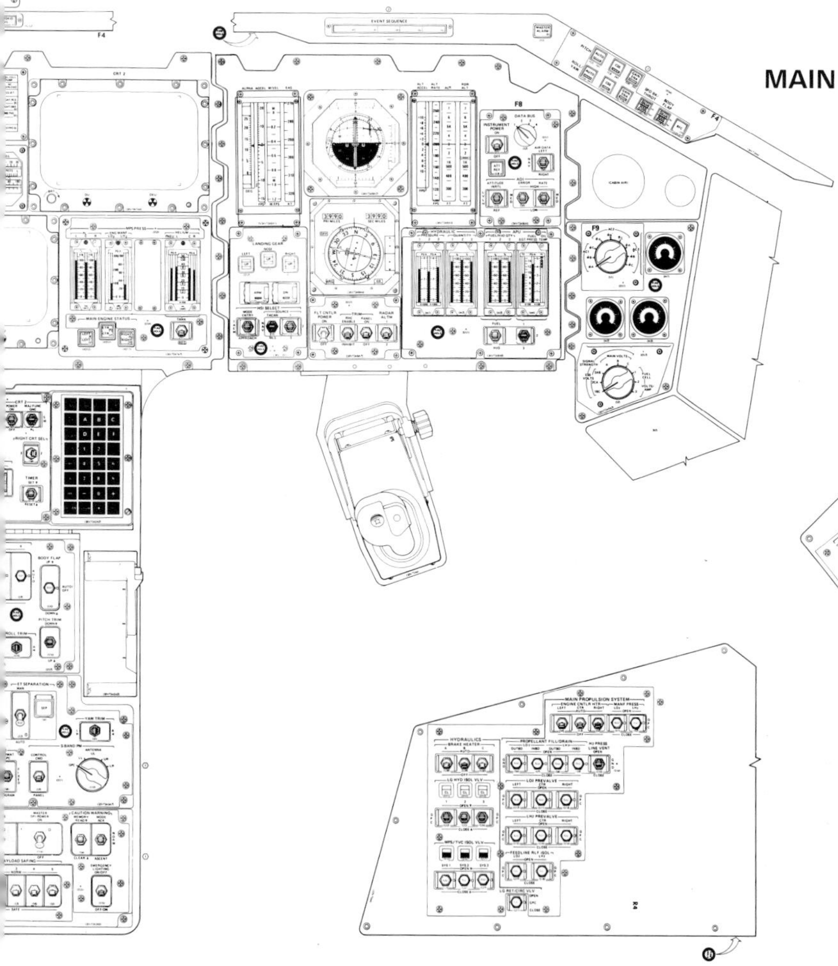

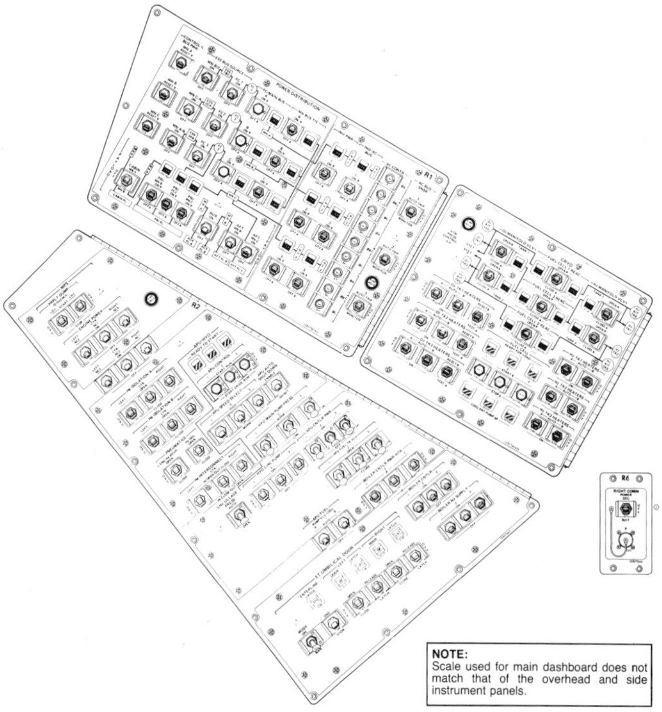

NOTE:
Scale used for main dashboard does not match that of the overhead and side instrument panels.

THE MAJOR MALFUNCTION

Twenty-four successful Shuttle missions took place before disaster struck. Critisisms were directed NASA's way from time to time in the (close-to) five-year period between the maiden Shuttle spaceflight, STS-1/*Columbia* in April 1981, and the last successful mission, 61-C/*Columbia* in January 1986, but for the most part the system performed as advertised.

In lambasting NASA, critics chose to overlook the fantastic technological leap the Shuttle represented. Here was a spacecraft demonstrably capable of making repeated visits to low-Earth orbit, being refurbished for reuse between missions, and landing airplane-style on a conventional runway instead of splashing-down in the ocean like the 'tin-can' craft of yore.

The Shuttle even demonstrated that it could capture faulty satellites in orbit, then return them to Earth for repair and relaunch – an awesome achievement in retrospect, and certainly a huge advance on previous space transportation systems.

Ironically, the Shuttle vehicle that was destroyed when tragedy finally struck was *Challenger,* which had made more spaceflights than any other Orbiter in the fleet. On board were seven people; five 'career astronauts' and two Payload Specialists.

One of these Payload Specialists was schoolteacher Christa McAuliffe, whose presence aboard *Challenger* as the first 'ordinary citizen' to make a spaceflight

attracted an uncommonly large degree of media attention to Shuttle mission 51-L.

There was live television coverage of the event, including a special video link-up with schools right across America...

Immediately after the disaster, President Reagan established a special accident investigation board under the chairmanship of former Nixon administration Secretary of State, William Rogers.

In their official report, published in June 1986, this investigative board – the Presidential Commission, or 'Roger's Commission' – described *Challenger's* final flight with these words:-

Flight of the Space Shuttle Challenger *on Mission 51-L began at 11:38am. Eastern Standard Time on January 28, 1986. It ended 73 seconds later in an explosive burn of hydrogen and oxygen propellants that destroyed the External Tank and exposed the Orbiter to severe aerodynamic loads that caused complete structural breakup. All seven crew members perished. The two Solid Rocket Boosters flew out of the fireball and were destroyed by the Air Force range safety officer 110 seconds after launch.*

The ambient air temperature at launch was 36 degrees Fahrenheit measured at ground level approximately 1,000 feet from the 51-L mission launch pad 39B. This temperature was 15 degrees colder than that of any previous launch.

The following description of the flight events is based on visual examination and image enhancement of film from NASA operated cameras and telemetry data transmitted from the Space Shuttle to ground stations. The last telemetry data from the Challenger *was received 73.618 seconds after launch.*

At 6.6 seconds before launch, the Challenger's *liquid fueled main engines were ignited in sequence and run up to full thrust while the entire Shuttle structure was bolted to the launch pad. Thrust of the main engines bends the Shuttle assembly forward from the bolts anchoring it to the pad. When the Shuttle assembly springs back to the vertical, the Solid Rocket Boosters' restraining bolts are explosively released.*

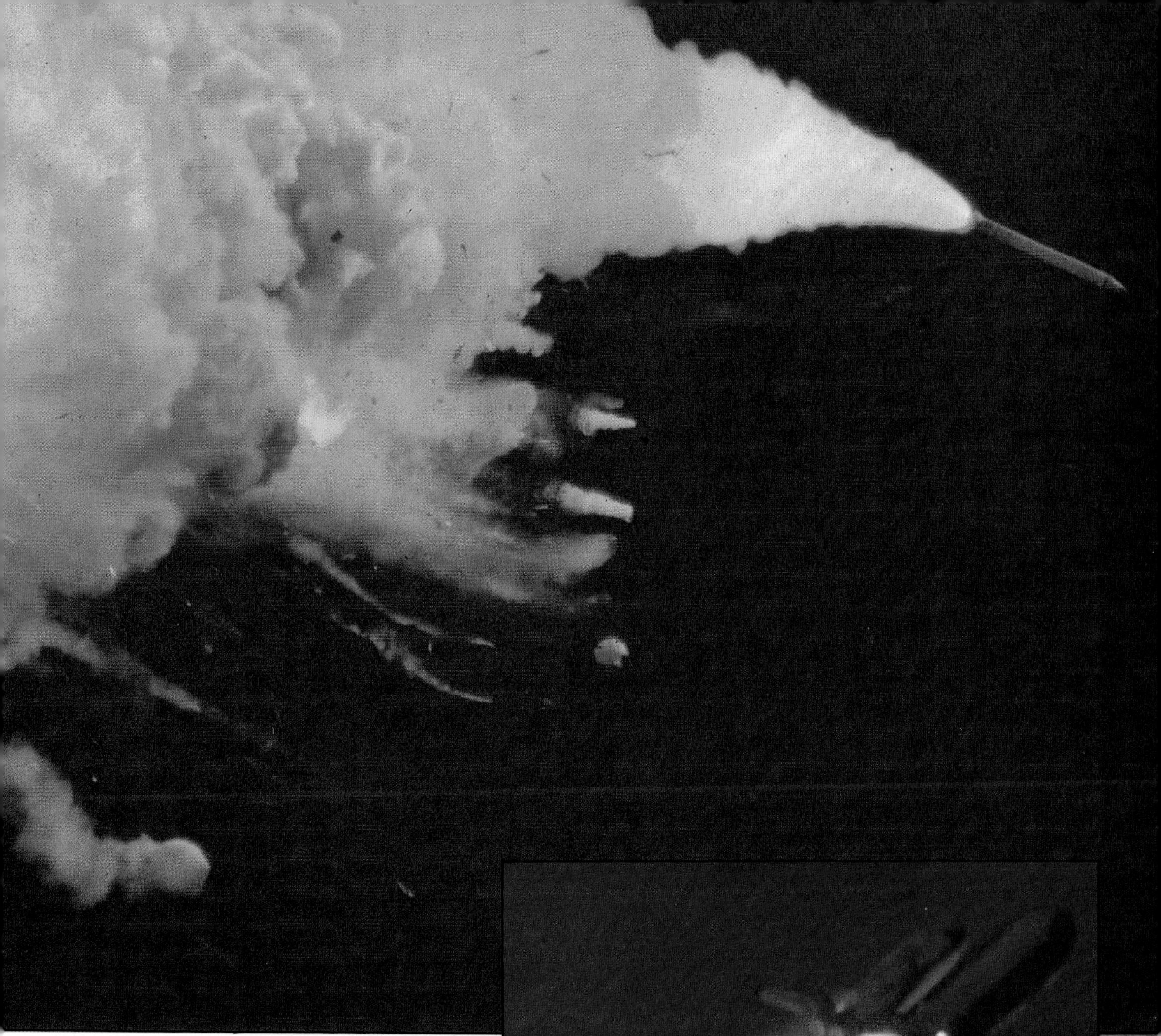

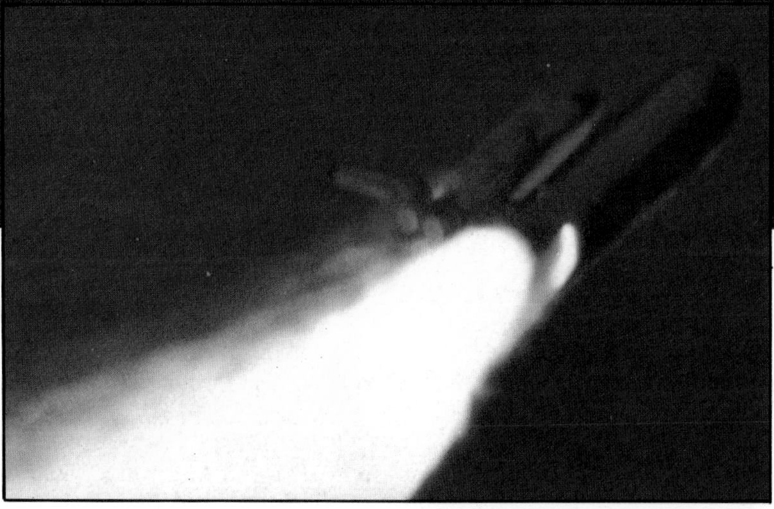

(*Top left*) Official crew insignia for the ill-fated 51-L/*Challenger* mission. Note the 'apple for the teacher' next to Christa McAuliffe's name.

(*Far left*) Graphic evidence that the weather conditions were extremely cold on 28 January 1986. This is part of Kennedy Space Center's Pad 39B, from which *Challenger* blasted away mere hours later.

(*Left*) Portent of disaster. Arrow points to tell-tale puff of thick black smoke from the eroded rubber O-ring seal in the aft field joint of the right-hand solid rocket booster at the moment of lift-off.

(*Right*) A tiny flickering flame indicates that the breach in the booster seal has somehow reopened.

(*Center right*) The flame grows in intensity.

(*Top right*) The left-hand SRB streaks away from the fireball after *Challenger's* disintegration.

During this pre-release "twang" motion, structural loads are stored in the assembled structure. These loads are released during the first few seconds of flight in a structural vibration mode at a frequency of about 3 cycles per second. The maximum structural loads on the aft field joints of the Solid Rocket Boosters occur during the "twang", exceeding even those of the maximum dynamic pressure period experienced later in flight.

Just after liftoff at .678 seconds into the flight, photographic data show a strong puff of gray smoke was spurting from the vicinity of the aft field joint on the right Solid Rocket Booster. The two 39B cameras that would have recorded the precise location of the puff were inoperative. Computer graphic analysis of film from other cameras indicated the initial smoke came from the 270 to 310-degree sector of the circumference of the aft field joint of the right Solid Rocket Booster. This area of the solid booster faces the External Tank. The vaporized material streaming from the joint indicated there was not complete sealing action within the joint.

Eight more distinctive puffs of increasingly blacker smoke were recorded between .836 and 2.500 seconds. The smoke appeared to puff upwards from the joint. While each smoke puff was being left behind by the upward flight of the Shuttle, the next fresh puff could be seen near the level of the joint. The multiple smoke puffs in this sequence occurred at about four times per second, approximating the frequency of the structural load dynamics and resultant joint flexing. Computer graphics applied to NASA photos from a variety of cameras in this sequence again placed the smoke puffs' origin in the 270-to 310-degree sector of the original smoke spurt.

As the Shuttle increased its upward velocity, it flew past the emerging and expanding smoke puffs. The last smoke was seen above the field joint at 2.733 seconds. At 3.375 seconds the last smoke was visible below the Solid Rocket Boosters and became indiscernible as it mixed with rocket plumes and surrounding atmosphere.

The black color and dense composition of the smoke puffs suggest that the grease, joint insulation and rubber O-rings in the joint seal were being burned and eroded by the hot propellant gases.

Launch sequence films from previous missions were examined in detail to determine if there were any prior indications of smoke of the color and composition that appeared during the first few seconds of the 51-L mission. None were found. Other vapors in this area were determined to be melting frost from the bottom of the External Tank or steam from the rocket exhaust in the pad's sound suppression water trays.

Shuttle main engines were throttled up to 104 percent of their rated thrust level, the Challenger executed a programmed roll maneuver and the engines were throttled back to 94 percent.

At approximately 37 seconds, Challenger encountered the first of several high-altitude wind shear conditions, which lasted until about 64 seconds. The wind shear created forces on the vehicle with relatively large fluctuations. These were immediately sensed and countered by the guidance, navigation and control system. Although flight 51-L loads exceeded prior experience in both yaw and pitch planes at certain instants, the maxima had been encountered on previous flights and were within design limits.

The steering system (thrust vector control) of the Solid Rocket Booster responded to all commands and wind shear effects. The wind shear caused the steering system to be more active than on any previous flight.

At 45 seconds into the flight, three bright flashes appeared downstream of the Challenger's right wing. Each flash lasted less than one-thirtieth of a second. Similar flashes have been seen on other flights. Another appearance of a seperate bright spot was diagnosed by film analysis to be a reflection of main engine exhaust on the Orbital Maneuvering System pods located at the upper rear section of the Orbiter. The flashes were unrelated to the later appearance of the flame plume from the right Solid Rocket Booster.

Both the Shuttle main engines and the solid rockets operated at reduced thrust approaching and passing through the area of maximum dynamic pressure of 720 pounds per square foot. Main engines had been throttled up to 104 percent thrust and the Solid Rocket Boosters were increasing their thrust when the first flickering flame appeared on the right Solid Rocket Booster in the area of the aft field joint. This first very small flame was detected on image enhan-

(Above) Veteran NASA photographer A.R. 'Pat' Patnesky captured this candid shot in the Mission Control Center at Houston at the instant *Challenger* exploded. Alarm and dismay are evident in the expressions of Flight Directors Lee Briscoe (foreground) and Jay Greene as they witness the scene at Cape Kennedy on the television screen at center.

(Below) Flight crew for the fatal Shuttle mission 51-L/*Challenger*. Left to right: El Onizuka, Mike Smith (Pilot), Christa McAuliffe, Dick Scobee (Commander), Greg Jarvis, Judy Resnik, and Ron McNair.

ced film at 58.788 seconds into the flight. It appeared to originate at about 305 degrees around the booster circumference at or near the aft field joint.

One film frame later from the same camera, the flame was visible without image enhancement. It grew into a continuous, well-defined plume at 59.262 seconds. At about the same time (60 seconds), telemetry showed a pressure differential between the chamber pressures in the right and left boosters. The right booster chamber pressure was lower, confirming the growing leak in the area of the field joint.

As the flame plume increased in size, it was deflected rearward by the aerodynamic slipstream and circumferentially by the protruding structure of the upper ring attaching the booster to the External Tank. These deflections directed the flame plume onto the surface of the External Tank. This sequence of flame spreading is confirmed by analysis of the recovered wreckage. The growing flame also impinged on the strut attaching the Solid Rocket Booster to the External Tank.

At about 62 seconds into the flight, the control system began to react to counter the forces caused by the plume and its effects. The left Solid Rocket Booster thrust vector control moved to counter the yaw caused by reduced thrust from the leaking right Solid Rocket Booster. During the next nine seconds, Space Shuttle control systems worked to correct anomalies in pitch and yaw rates.

The first visual indication that swirling flame from the right Solid Rocket Booster breached the External Tank was at 64.660 seconds when there was an abrupt change in the shape and color of the plume. This indicated that it was mixing with leaking hydrogen from the External Tank. Telemetered changes in the hydrogen tank pressurization confirmed the leak. Within 45 milliseconds of the breach of the External Tank, a bright sustained glow developed on the black-tiled underside of the Challenger *between it and the External Tank.*

Beginning at about 72 seconds, a series of events occurred extremely rapidly that terminated the flight. Telemetered data indicate a wide variety of flight system actions that support the visual evidence of the photos as the Shuttle struggled futilely against the forces that were destroying it.

At about 72.20 seconds the lower strut linking the Solid Rocket Booster and the External Tank was severed or pulled away from the weakened hydrogen tank permitting the right Solid Rocket Booster to rotate around the upper attachment strut. This rotation is indicated by divergent yaw and pitch rates between the left and right Solid Rocket Boosters.

At 73.124 seconds, a circumferential white vapor pattern was observed blooming from the side of the External Tank bottom dome. This was the beginning of the structural failure of the hydrogen tank that culminated in the entire aft dome dropping away. This released massive amounts of liquid hydrogen from the tank and created a sudden forward thrust of about 2.8 million pounds, pushing the hydrogen tank upward into the intertank structure. At about the same time, the rotating right Solid Rocket Booster impacted the intertank structure and the lower part of the liquid oxygen tank. These structures failed at 73.137 seconds as evidenced by the white vapors appearing in the intertank region.

Within milliseconds there was a massive, almost explosive, burning of the hydrogen streaming from the failed tank bottom and the liquid oxygen breach in the area of the intertank.

At this point in its trajectory, while traveling at a Mach number of 1.92 at an altitude of 46,000 feet, the Challenger *was totally enveloped in the explosive burn. The* Challenger's *reaction control system ruptured and a hypergolic burn of its propellants occurred as it exited the oxygen-hydrogen flames. The reddish brown colors of the hypergolic fuel burn are visible on the edge of the main fireball. The Orbiter, under severe aerodynamic loads, broke into several large sections which emerged from the fireball. Seperate sections that can be identified on film include the main engine/tail section with the engines still burning, one wing of the Orbiter, and the forward fuselage trailing a mass of umbilical lines pulled loose from the payload bay.*

Evidence in the recovered wreckage from the 51-L mission hardware supports this final sequence of events.

The wreckage recovered from the Atlantic Ocean in subsequent weeks, during the largest water salvage operation ever mounted, bore grim testimony to the mechanism of the tragedy.

Significant pieces of the Orbiter vehicle itself were recovered, including all three SSME main engines, the forward fuselage section containing the remains of the crew, the right-hand (starboard) inboard and outboard elevon control surfaces, a large portion of the starboard wing, a lower portion of the vertical tailfin structure, three rudder/speed-brake panels, and portions of the mid-fuselage side walls from both the port and starboard sides.

The crew module wreckage was found submerged in about 90 feet of water, concentrated in an area about 20 feet by 80 feet. Portions of the forward fuselage outer shell structure were found among the pieces of crew module recovered.

There was no evidence of an internal explosion, heat or fire damage on the forward fuselage/crew module pieces. The cabin section was distintegrated, with the heaviest fragmentation and crash damage on the left side. The fractures examined by the investigators were typical of overload breaks and appeared to be the result of high forces generated by impact with the surface of the water, rather than by the midair explosion itself.

It was later conceded that some, if not all of the crewmembers, may have been alive until the moment the cabin section hit the water, although they were almost certainly unconscious.

The hulks of all three main engines were recovered off the Florida coast in about 85 feet of water. The engines themselves were still attached to their load-bearing thrust structure and all associated ancillary parts were situated close-by. All engine gimbal bearings had failed, again apparently due to the force of the impact with the water.

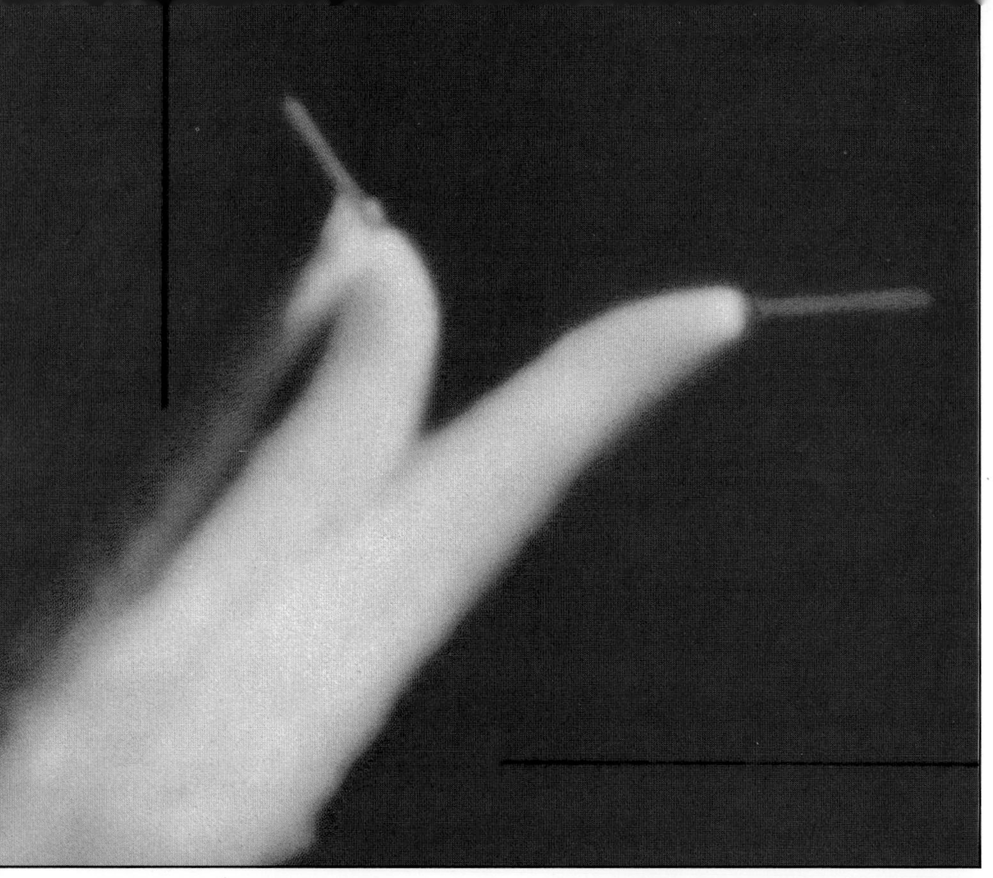

(Left) Close-up shot showing the way the two solid rocket boosters crossed over in mid-air soon after *Challenger's* disintegration.

(Below) Three identifiable pieces of debris are visible in this photo, taken by a NASA tracking camera. The top arrow points at one of the wings, while the center arrow indicates the location of *Challenger's* cluster of three main engines (SSMEs), which are still operating here. The bottom arrow shows that the crew module with the seven astronauts inside emerged intact from the fireball.

(Right) Such was the violence of the explosion that the accident was visible from space. This extraordinary image was captured by a GOES weather satellite orbiting the Earth at an altitude of 22,250 miles. Arrows indicate the locations of the white smoke trail *Challenger* left behind as she ascended from the Cape, and of the smoke cloud from the explosion itself. The feature just above the dateline is Lake Ocheechobee, which – to give some impression of scale – measures approximately 35 miles by 30 miles.

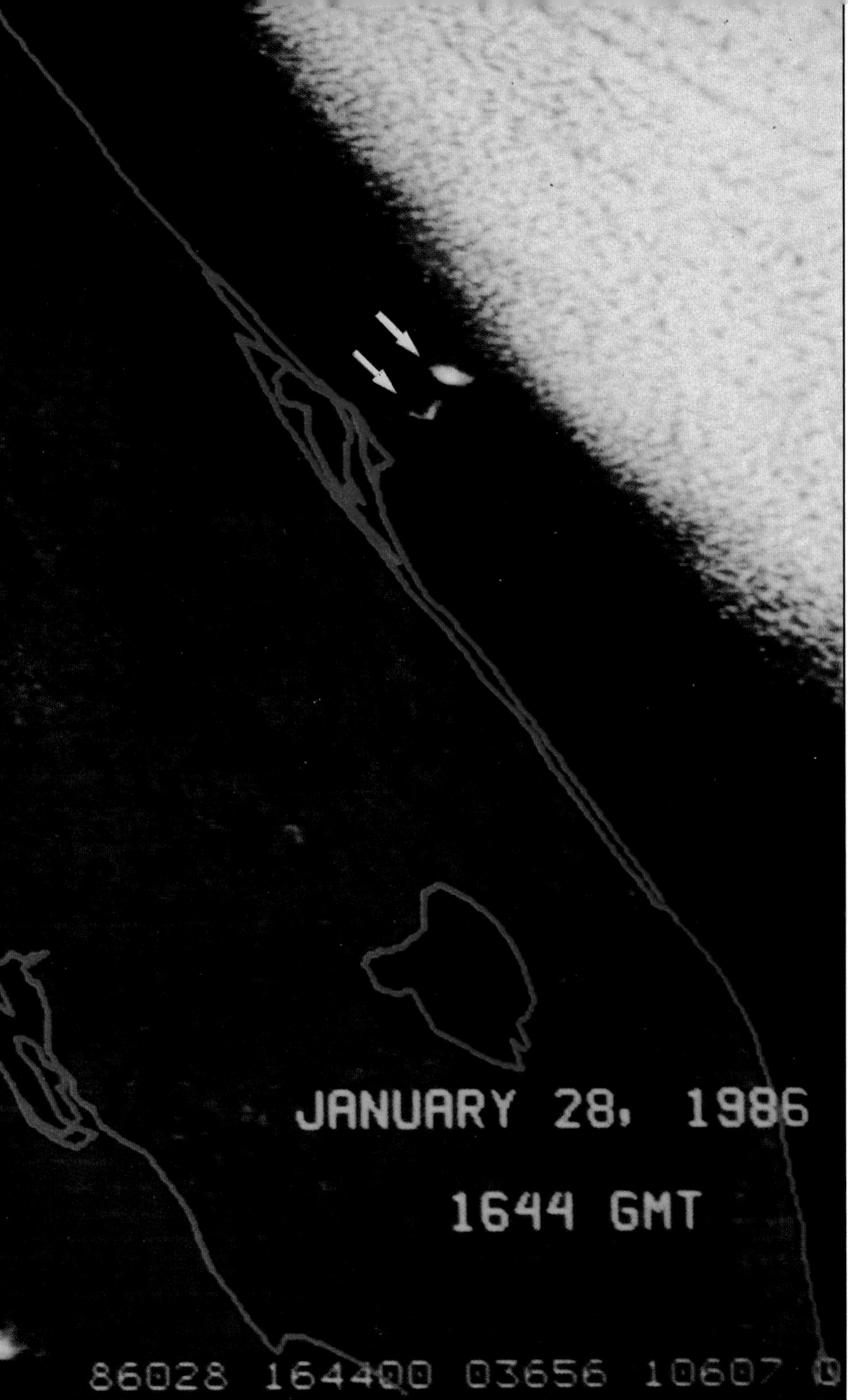

All metallic surfaces were damaged by marine life, except titianium parts or those parts that were buried just under the ocean floor.

Sections of the main propulsion system fuel and liquid oxygen feedlines were recovered, as well as the disconnect assembly between the Orbiter and the External Tank, which was still in the 'mated' configuration. The main engine controllers for two of the engines were recovered.

One controller was broken open on one side and both were severely corroded and damaged by marine life. Incredibly, though, once both units were disassembled and the memory units flushed with deionized water, then dried and vacuum-baked, data stored in these units was retrieved to assist the investigation...

The most telling evidence was found in the vicinity of a joint between two segments of the right-hand solid rocket booster. As expected, in summarizing all the available evidence, the Presidential Commission came out firmly in favor of the theory that the loss of *Challenger* was caused by a failure of a rubber O-ring seal within that joint, which allowed searingly-hot combustion gases from within to burn a hole through the outer casing of the booster and sever one of the struts attaching the booster to the External Tank.

The booster then twisted free of a second attachment strut at its base and pivoted about its upper attachment point, allowing the booster's conical tip to puncture the liquid oxygen vessel housed in the upper end of the External Tank.

This triggered an enormous explosion, which destabilized the craft and allowed aerodynamic forces at nearly 1,500 mph to literally tear the whole assembly apart.

Contrary to popular reports, the 111 tons of recovered *Challenger* wreckage was not permanently buried. It now lies deep underground in two disused Minuteman missile silo complexes in a quiet corner of Cape Kennedy. If ever there was any need to bring some out, it could be done.

The silo complexes had deteriorated badly after some 17 years of neglect. Therefore, the operation to prepare the site as a final resting place was a tricky one. There was extensive photo-documentation by NASA lensmen, and the whole operation was conducted under an tight security cordon, with around-the-clock surveillance to deter people with a morbid interest in the wreckage, like troublesome onlookers – or even souvenir hunters.

Elliot Kicklighter was in charge of the debris entomoment operation, and I spoke to him at Kennedy Space Center about the problems involved.

Born in Savannah, Georgia in 1944, Kicklighter joined NASA at Kennedy Space Center in June 1967, a few short months after another tragedy; the Apollo 1 pad fire which claimed the lives of astronauts Grissom, White and Chaffee.

Today, Kicklighter is working under the command of veteran Shuttle astronaut Robert Crippen, who moved to Kennedy Space Center in June 1987 to play a prominent role in the Shuttle rehabilitation program. In the immediate post-51-L timeframe, Kicklighter had shouldered the onerous duties of *Challenger* 'Debris Manager'.

The $100-million operation to actually recover the wreckage from the Atlantic Ocean was commanded by Colonel Edward O'Connor of the U.S. Air Force, who has since retired. Elliot Kicklighter and his approximately 15-strong team then had the responsibility of handling all the hardware when it was brought in by the recovery ships to the dock.

After being off-loaded from the fleet of recovery vessels at Port Canaveral, the wreckage was laid out for investigation purposes in temporary storage areas while the National Transportation Safety Board (NTSB) team led by Terry Armentrout conducted the painstaking study effort under the watchful eye of the Rogers Commission appointed by President Reagan.

The 111 tons of wreckage recovered from the ocean represented approximately 35 percent of the Orbiter vehicle, 50 percent of the solid rocket boosters and about 50 percent of the big orange External Tank. There were also significant chunks of the TRDS-B and Spartan-Halley satellite payloads that had been housed in *Challenger's* cargo bay.

The bulk of the wreckage from the Orbiter and the External Tank was stored in one 13,000-square-foot corner of a gigantic new building called the Kennedy Space Center Logistics Facility, which has since reverted to its proper use as an Orbiter spare parts storage facility and ground support equipment storage zone.

Meanwhile, recovered solid rocket motor components were placed in Hanger 'O' on nearby Cape Canaveral Air Force Station (CCAFS).

That work went on from about February to August of 1986. To assist the official inves-

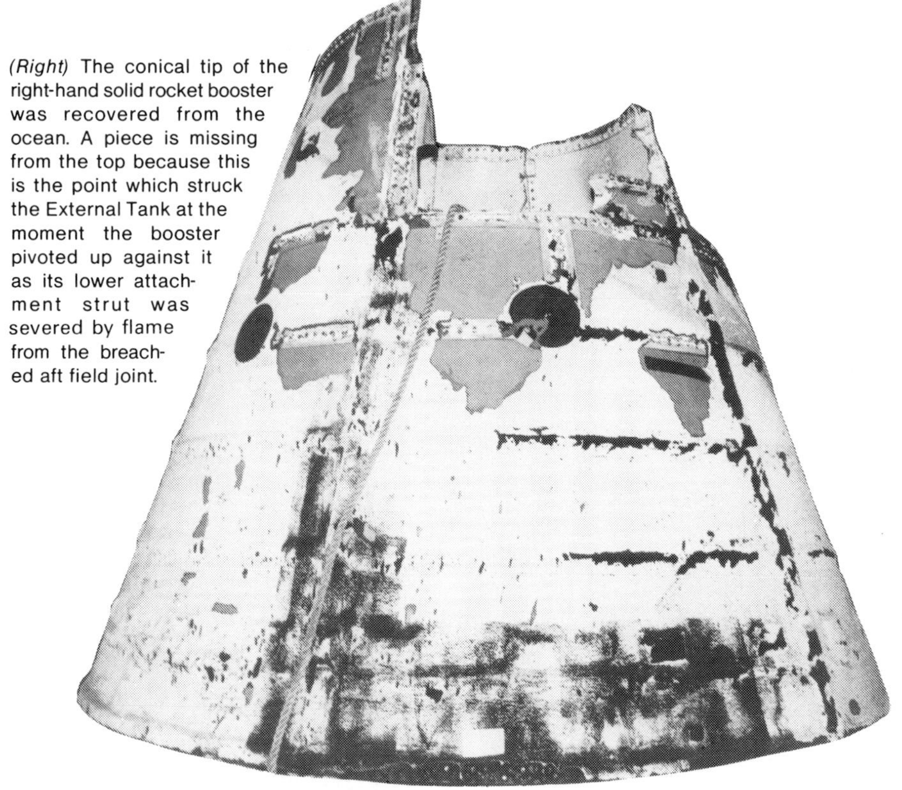

(Right) The conical tip of the right-hand solid rocket booster was recovered from the ocean. A piece is missing from the top because this is the point which struck the External Tank at the moment the booster pivoted up against it as its lower attachment strut was severed by flame from the breached aft field joint.

THE LAST CONTACT

INCREDIBLY, the flight deck voice recorder was located and recovered from the water, and a transcript of the 51-L crew's final words was produced by a team led by former Skylab astronaut Joe Kerwin. The 'T' symbol denotes the moment of lift-off. Therefore, 'T-6' means six seconds before lift-off. 'T+40' means 40 seconds after lift-off.

T−6 – Scobee: *"Three at a hundred."* (main engines)
T0 – Resnik: *"Alll riiight!"*
T+1 – Smith: *"Here we go."* (vehicle motion)
T+7 – Scobee: *"Houston, Challenger roll program."* (call to controllers)
T+11 – Smith: *"Go, you mother!"*
T+14 – Resnik: *"LVLH."* (reminder for cockpit switch change)
T+15 – Resnik: *"Shit hot!"*
T+16 – Scobee: *"Ohhhkaaay!"*
T+19 – Smith: *"Looks like we got a lotta wind here today."*
T+20 – Scobee: *"Yeah."*
T+22 – Scobee: *"It's a little hard to see out my window here."*
T+28 – Smith: *"There's 10,000 and Mach point five."*
T+30 – Garble.
T+35 – Scobee: *"Point nine."*
T+40 – Smith: *"There's Mach one."*
T+41 – Scobee: *"Going through 19,000."*
T+43 – Scobee: *"Okay, we're throttling down."*
T+57 – Scobee: *"Throttling up."*
T+58 – Smith: *"Throttle up."*
T+59 – Scobee: *"Roger."*
T+60 – Smith: *"Feel that mother go."*
T+60 – Uncertain: *"Woooohoooo!"*
T+62 – Smith: *"35,000 going through one point five."*
T+65 – Scobee: *"Reading 486 on mine."* (routine airspeed check)
T+67 – Smith: *"Yep, that's what I've got too."*
T+70 – Scobee: *"Roger, go at throttle up."* (responding to ground call)
T+73 – Smith: *"Uh-oh."*
T+73 – Loss of all data

tigators, Kicklighter and his team produced a rough *Challenger* 'mockup' by arranging the pieces of debris in their original positions relative to one another.

The wreckage remained in this 'footprint' configuration all the way through to the end of the year, because once the investigations were over and the reports were written, NASA got down to some internal engineering evaluation work for data-retrieval purposes, analysing the hardware.

In January 1987, almost one year to the day after the accident, this data-retrieval operation was completed and a decision taken to disassemble the 'footprint' and move the wreckage to a permanent storage area.

Surprisingly little equipment was used to move the pieces of debris about; a 65-ton-capacity crane, two fork-lifts, and two 60-foot-long flatbed trucks to finally move the wreckage from the temporary storage sites to the permanent facility.

Challenger's final resting place can perhaps be characterized as bizarre, but practical. It was Elliot Kicklighter's team that modified the two disused Minuteman missile complexes situated on an isolated region of Cape Canaveral Air Force Station, well away from the public gaze.

The chosen site consists of two silos (Complexes 31 and 32) and four vault-like underground equipment rooms. The latter still housed a considerable amount of electronic equipment associated with missile operations, which Kicklighter's team had to remove before the *Challenger* wreckage could be transported in.

There were other important tasks to perform as various modifications were put in hand to transform the underground equipment rooms into permanent storage vaults. They were in a bad state of repair after years of neglect. Although the atmosphere was to be left ambient, with no environmental control, if the precious debris was to withstand the onslaught of time, the underground facilities had at least to be made watertight.

Back in the late-sixties, the silos and their equipment rooms had been bone-dry, but now the damp was everywhere. A search of old Kennedy Space Center records revealed how, ten years before, a burst pipe had immersed the floor of one silo under several feet of water.

All the waterlines were capped-off before any of *Challenger's* debris was moved in.

Because the debris was recovered from salt water, some of it required corrosion-proofing to preserve it. For example, corrosion protection was applied to the critical tang and clevis (tongue and groove) area of the aft field joint of the right-hand solid rocket booster, where the original cause of the accident lay.

In addition, special measures were taken to protect *Challenger's* cockpit voice-recorders, which were among the last components to be placed in the silos. Soon after recovery they had been sent to Johnson Space Center in Houston for painstaking restoration by the IBM company and analysis by a team led by ex-Skylab astronaut, Dr Joe Kerwin (see cockpit voice-recorder transcript at the bottom left-hand corner of this page).

"The wreckage was placed in the vaults in a very organized manner", says Elliot Kicklighter. *"We compartmentalized the wreckage, so that all of the Orbiter pieces went into one area, all of the External Tank components were put together in another area, and all the solid rocket motor pieces went somewhere else.*

"I guess we just used a common-sense approach to make it all fit into the available areas. Primarily, the larger components were put put in first, and anything we felt would be of any significance in the future was left in an accessible area.

"It was all logged-in by our quality-control personnel here at the Cape, and the entries put into official debris logbooks. These record precisely where each component is stored."

Some of the pieces were so big that they had to be cut before they would fit into the silos. For example, *Challenger's* starboard wing was virtually intact and had to be cut in two.

Elliot Kicklighter: *"Some people expressed concerns when they heard we were planning on doing that, but the cutting was done in a very orderly manner by seeing to it that photographs were taken of the wreckage before and after cutting, and that kind of thing.*

"There was never anything really disturbed or modified. The cutting we did was straight line-type of thing and was very minimal. It was also well-documented, to record the status of any particular piece or component for the purposes of any kind of future investigation."

The largest pieces of debris measured about twenty or thirty feet across. In addition to the Orbiter's starboard wing, components requiring cutting included some

(Right) The 'Stars and Stripes' sees an ignominious end. Although the debris is stored underground, it could be accessed for re-examination by a future investigative board should this ever prove necessary.

(Below) A section of *Challenger's* fuselage side being lowered into its final resting place at Cape Canaveral.

chunks of the External Tank and the solid rocket boosters.

The very methodical manner in which the pieces were cut allayed any concerns that the severing of components might irretrievably distort the 'evidence' available to any future investigative team.

It would be a fairly involved operation for anyone to get back in, but such access could be facilitated if the need so arises. Future investigators could go in and retrieve components after a few days of clearing work was completed, but there are no plans for periodic opening-up of the vaults to check the condition of the stored wreckage.

"We plan on leaving it indefinitely', says Elliot Kicklighter. *"Hopefully we won't have to go back in, because all of the engineering investigations that we know of have been done."*

As aforementioned, most of the debris was entombed in the February/August 1987 timeframe, and the operation wound right down. However, as recently as the first week of December 1987, when I interviewed Elliot Kicklighter, some of the components retained for further inspection were only just being finally committed to the ground.

By that stage, the only pieces yet to be laid in place were the nose and mainwheel landing gears, which were being held by Rockwell International at Downey, California for research and development work associated with the Orbiter braking system, and some key pieces of the solid rocket motor cases and internal insulation, which were in the hands of manufacturers Morton Thiokol at Brigham City, Utah undergoing further engineering evaluation.

When these few poignant remains were finally laid in position, the last of the huge 10-ton concrete caps was secured with long steel rods and welded down over its underground vault to prevent unwelcome access.

Challenger is now sealed-in, resting, perhaps for all time.

(Below) This piece of debris from the starboard wing bears the titling *Challenger*, barely visible through the layer of corrosion. It is a grim symbol of a sad chapter in space history.

BACK TO STRENGTH

Challenger's demise heralded the end of an era for America's manned space program. Never again could NASA's publicity handouts be taken at face-value. From here on in, the agency's claims as to its capabilities would be subjected to intense public and media scrutiny.

Almost immediately after the Roger's Commission began its formal investigation into the causes of the *Challenger* disaster, NASA and its team of contractors stepped-up efforts to improve the Shuttle system, determined to prevent a repetition of the events that brought about its fall from grace. The tasks they undertook can be loosely divided into three categories.

First: rectifying the design faults that were the direct cause of the accident – namely, redesigning the joints between the booster rocket segments.

Second: identifying other components which might fail in flight with similarly catastrophic results, then affecting appropriate remedial action. Potential weak-spots were found in various areas; not just in the boosters, but in the Orbiter vehicle itself.

Third: instigating sweeping changes to NASA's management structure in general and, in particular, the administration of the Shuttle program itself. It was clear that a fundamental breakdown in communications through the line of command had prevented concerns about weak-spots in the Shuttle's design from being relayed to a higher authority for attention.

The most tangible facet of these three elements of the post-*Challenger* phase was undoubtedly the modification work that got the three surviving Orbiters back into a revised condition of flightworthiness for the resumption of Shuttle missions. A few short miles from the point where *Challenger* was torn into a thousand pieces, *Discovery*, *Atlantis* and *Columbia* were being prepared for their next missions.

Discovery was being prepared for transfer to Vandenberg Air Force Base in California, where she was to have inaugurated Shuttle launches into polar orbit from the specially-modified facilities there; *Atlantis* was ensconced in High Bay 1 of the Orbiter Processing Facility (OPF), but was within a week of being rolled into the nearby VAB for mating with her External Tank and SRBs in preparation for launching the Galileo probe to Jupiter in May; and *Columbia* had just been flown back from Edwards, California atop the Boeing 747 SCA following the 61-C mission.

(Below) **A full-scale, full-duration (120-second) test firing of a redesigned Shuttle solid rocket booster at the Morton Thiokol company's Utah plant.**

In the immediate aftermath of the *Challenger* disaster, the official word from NASA headquarters in Washington D.C. was that the agency would endeavor to maintain the original launching schedule. As the full implications of the tragedy gradually became apparent, however, it was clear that preparations for further flights would be severely disrupted and that the technical specification of the three surviving Orbiters would have to be considerably upgraded before they were considered safe enough to enter space again.

Definition and development of these modifications took many months, and when they were eventually implemented, priority was given to the Orbiter destined to make the first post-*Challenger* mission, *Discovery*. The three most significant modifications incorporated into each of the Orbiter vehicles were the installation of a crew escape system, the fitment of new-specification valves where the feed lines from the External Tank joined with the Orbiter, and the incorporation of upgraded thermal protection immediately aft of the spacecraft's carbon-carbon nosecap.

The Shuttle's crew escape system would not save the astronauts from every type of emergency, but at least it was an improvement on the almost non-existent set of options available previously. The new system has been designed for use in controlled, gliding flight, at altitudes ranging from 24,000 feet down to about 11,000 feet following failures and difficulties during ascent and re-entry where landing at a suitable airfield cannot be achieved.

Previous procedures for ascent aborts, where no landing site could be reached, required the Orbiter to ditch in the ocean, a condition which had since been shown by structural analysis to be non-survivable.

A 127-inch-long telescopic pole device was the primary element of the new system. Prior to its adoption, it was tested by volunteer parachutists from the U.S. Navy, who made a total of 66 jumps from a hatch-like opening in a Lockheed C-141 airplane over Edwards, California. To aid speedy deployment of the lightweight aluminium pole, pyrotechnic devices were fitted to the crew module hatch to enable it to be blown clear. This hatch jettison system would also be highly beneficial for certain other emergencies arising after touchdown (a situation aided by the incorporation of an inflatable slide similar to those provided on commercial airliners).

Escaping astronauts sliding down the

telescopic pole are guided by a lanyard attached to their parachute harness, in order to ensure that they fall clear of potentially lethal protuberances, such as the leading edge of the Orbiter's port wing and bulbous left-hand OMS pod.

(An alternative crew escape system – a 'tractor-rocket' device that would physically pull the astronauts through the jettisoned hatch and clear of the wing and OMS pod – was dropped due to its inherent complexity and storage hazards).

Incorporation of an in-flight escape system completed a comprehensive 'package' of crew safety measures which included a partial-pressure suit comprising oxygen equipment, a parachute, a life-raft and survival equipment for each crewmember. The partial-pressure suits would not save the astronauts in the event of a complete cabin depressurization, but they would facilitate their survival in the event of a short-duration partial depressurization

The second major modification to the Orbiter was incorporation of a new-specification version of the '17-inch disconnect' system: comprising two giant feeder-valves set into the Orbiter's belly, which connect with similar devices on the External Tank to carry the flow of propellant to the three main engines situated in the aft end of the Orbiter.

Although there had never been any sign of an imminent malfunction in this particular region, tests had indicated that under certain (highly unlikely) circumstances, the existing 17-inch disconnect valve design could go into a 'flutter' regime and effectively shut off the supply of propellant, resulting in total failure of the main engines. The redesigned system would prevent such a situation.

The third major Orbiter modification was not incorporated until after the Shuttle's return to flight operations. It was the installation of an aft extension to the spacecraft's reinforced carbon-carbon (RCC) nosecap. Although no significant damage had been noted in this vicinity subsequent to previous flights, tests had shown the spacecraft's 'chin' area, from the trailing edge of the nosecap to the forward edge

***(Above)* Rick Hauck, Commander of the first post-*Challenger* mission.**

***(Below)* Test firing of the (now abandoned) tractor-rocket escape system.**

***(Below)* A demonstration of the telescopic-pole escape system.**

***(Above) Columbia* receiving tender loving care in the Vehicle Assembly Building (VAB) at Kennedy Space Center prior to the Shuttle fleet's return to flight status in the wake of the *Challenger* disaster.**

of the nose landing-gear cavity, to be extremely temperature-critical during re-entry.

Vought, of Dallas, Texas produced the new 'chin' panel. The company also manufacture the Orbiter's nosecap and wing leading-edge panels from RCC, the Shuttle's most capable heat-resistant material.

As well as the very considerable number of modifications made to the three Orbiter vehicles, personnel at the Brigham City, Utah plant of the Morton Thiokol Corporation devoted a tremendous amount of time ensuring that the Shuttle's solid rocket boosters – the essential source of *Challenger's* demise – were put to rights. A series of full-duration (120-second) test firings of the SRBs in the horizontal position, along with numerous short-duration firings of 'stunted' partial-segment SRBs, verified that a variety of fundamental design changes had indeed made the operation of these potent boosters safer and more reliable.

Full details of the original design of the SRBs can be found in the chapter entitled *Launch!*, and information about the SRB joint failure that caused the *Challenger* disaster is contained in the chapter entitled *The Major Malfunction*. To summarize: the pencil-like body of each SRB is composed of eleven cylindrical segments, which are joined to form four primary 'casting segments' at Morton Thiokol's Utah factory. These are stacked one atop the other on the Mobile Launcher Platform inside the VAB at Kennedy Space Center at an early stage in the launch campaign. The joints between the casting segments are called field-joints because these segments are mated 'in the field', as opposed to within the manufacture's facility.

The Rogers Commission determined that there were a number of key factors in the 51-L SRB failure, the foremost being the reaction of the rubber O-rings located between the field-joints in the right-hand SRB to the very low temperatures at the Cape on the launch morning. The fatal fault lay in the inability of the original O-ring design to cope with such low temperatures: a problem compounded by the tendency of the joints to spread, rather than seal, under the pressure of ignition.

Furthermore, there were also concerns about the existing SRB design's susceptibility to water intrusion into the joints, because at very low temperatures this would result in water turning to ice and seriously impeding the joint's ability to seal (or 'seat') correctly against a possible onslaught from the searingly hot gases within.

Further concerns stemmed from the fact that there were inadequate means of inspecting certain critical components of the rocket motor after it had been assembled for flight.

The upshot of an intensive investigation was not only the definition of a requirement to develop O-ring seals from a new type of material, but also a thorough evaluation of the whole rocket motor design: not just the notorious field-joints. In time, Morton Thiokol introduced what is perhaps the most significant single modification to the original SRB design. This is the addition of a so-called 'capture feature' behind the original 'tang-and-clevis' (or tongue-and-groove) fittings at the three field-joints.

Essentially, instead of having a tang which goes into a clevis, there are now *two*

(Right) A tile technician helps upgrade *Columbia* to her 'return to flight' specification.

clevises designed such that one leg of one clevis fits into the opening of the other clevis, forming an interlocking set of clevises. This new configuration ensures that 'casting segments' stack atop their partners in an manner analogous to a metal lid snapping tightly shut onto a paint can: an action which forces the two elements together and locks them into place so that the O-ring seals located between them cannot now move with respect to the metal parts.

Such movement (known as 'joint rotation') was found to be a contributory cause of the *Challenger* disaster. The new twin-clevis/capture feature design reduces joint rotation by an impressive 500 percent.

In another significant modification to the original SRB design, there are now three O-rings, not two. There are also now good-quality viewing ports to aid inspection of the field-joints, and electrical heater elements wrapped around the joints to ensure that their temperatures never drop below a predetermined safety level.

As well as devising the field-joint fix, Morton Thiokol introduced modifications to the internal insulation which lines each booster segment. The Thiokol engineers also found that they could make improvements to the area where the huge aft nozzle is mounted on the back of the booster case, at the so-called case-to-nozzle joint.

The final area where major changes were made was in the aft nozzle itself. The nozzle is primarily a shell made up of metal parts and covered with an ablative material (ablation is a process of discarding heat by shedding bits of a heavy, resinous coating protecting vulnerable parts from extreme heat levels). Morton Thiokol found a number of ways to improve the design of the nozzle – both in the plastic ablative material components and in the bolted-together metal parts from which the nozzle is composed – by putting in extra seals to provide a more assured sealing capability.

Yet another improvement made in the aft nozzles was the provision for pressure-testing of the seals after assembly to ensure that they aren't leaking.

These booster modifications and a host of large and small changes made to the specification of the three surviving Orbiters, acting in concert with many thousands of smaller improvements made to other elements of the Shuttle system and its management structure assured a safe return to flight status with the successful launch of STS-26/*Discovery* in September 1988. Further improvements have been introduced since that time. Late in 1987, NASA announced that the Orbiter's permitted end-of-mission landing weight would be increased from 211,000 pounds to 230,000 pounds, after a detailed structural analysis and review of the forces encountered by the Orbiter during landing indicated that the reusable spacecraft could safety carry such a load.

(Right) A technician at work near *Discovery's* 'chin' area.

Speaking of this development, Rear Admiral Dick Truly, a former astronaut and now head of the Space Shuttle program, said: "The total Space Shuttle performance capability requires a balance between lift-to-orbit and the allowable return weight during re-entry and landing. This new capability will improve this balance and add considerable flexibility and efficiency to our Space Transportation System."

More recently, high-efficiency carbon brakes have found their way onto the Orbiters, and drag-chutes are to come into use to reduce the strain on the landing gears. The Shuttle also totes new computers, twice as fast as the original units, yet half the size and weight.

In May 1991, a second Boeing 747 Shuttle Carrier Aircraft (SCA) entered NASA service, increasing ferrying capability and eliminating a potential 'single-point failure' in the Shuttle system. Bear in mind that, if the original SCA had crashed or been rendered inoperative for some other reason, the entire Shuttle fleet would have had to be grounded for lack of a means of transporting the Orbiters back from their landing sites to the Kennedy Space Center.

The new SCA's first task was to ferry NASA's new Orbiter vehicle, *Endeavour* (OV-105) from manufacturer Rockwell International's Palmdale, California plant to Kennedy Space Center. *Endeavour*, the *Challenger*-replacement Orbiter, will make its first space flight in April 1992.

As the relentless process of reinforcing weak-spots in the Shuttle system has progressed, the reusable spacecraft has gradually adopted a new position in America's space program. It is now firmly established as the manned element of a so-called 'mixed-fleet' philosophy, sharing the workload with a wide range of unmanned, expendable launch vehicles: McDonnell Douglas Deltas, General Dynamics Atlas-Centaurs, Martin Marietta Titans and LTV Scouts. Speaking in early-1988 at the formal handover of the first of 23 Titan 4 boosters ordered by the U.S. Air Force, then-Air Force Secretary Edward C. 'Pete' Aldridge said that a "tragic error" had been made in the 1970s when the USAF placed reliance on the Shuttle as its sole launch vehicle.

"We have paid, and will continue to pay, dearly for that error. In fact, it will cost the Department of Defense about $10 billion to restore a balance between the Shuttle and expendable launch vehicles to recover from that space policy mistake". Over the ensuing years, the role of the Shuttle in military space operations has steadily diminished, and the top-secret security classification for such flights was lifted for good in 1991.

***(Right)* The STS-41/*Discovery* crew signifies hearty approval of the reborn Shuttle program.**

8 PERSONS
OR2000 LBS.

FINANCIAL YEARS

Any outline of the costs involved in mounting a project as complex as the Space Shuttle's development program is complicated by questions like; *"Are the figures based on a single launch, or averaged over an Orbiter vehicle's total lifespan? Do we talk 1969 dollars, 1974 dollars or 1988 dollars?"* On top of that, inflation – not just actual expenditures – can push the figures up to misleading levels.

For these reasons, the figures quoted here should be regarded as simplifications, but they are all officially-quoted NASA estimates.

In Financial Year (FY) 1988 dollars, NASA estimated the cost of an Orbiter vehicle – including engines and government-furnished items such as the RMS robotic arm, the galley and the closed-circuit television system – to be approximately $1.7 billion. This equates to approximately $658 million in FY 1971 dollars. The original estimate for an Orbiter was $250 million in FY 1971 dollars, which was later revised to $350-$400 million FY 1971 dollars, based on the reduction of fleet size from five to four Orbiters and a schedule stretchout.

NASA's original estimate for the Orbiter *Columbia's* thermal protection system (TPS) of tiles and other materials was approximately $200 million in FY 1971 dollars, which included development, manufacturing and installation. The actual cost for *Columbia's* TPS was $315 million.

Empty, a Shuttle External Tank costs $13 million FY 1988 dollars. External Tank development costs were $597 million, as compared to the equivalent FY 1971 estimate of $392 million. Comparable figures for the solid rocket boosters are $564 million (equivalent $366 million FY 1971), and for the Space Shuttle Main Engines, or SSMEs, $1,403 million (equivalent $947 million FY 1971).

Assessing the Space Shuttle's fuel bill is an interesting exercise. Liquid oxygen for one flight (141,000 gallons) costs 38 cents a gallon, or $53,580. Liquid hydrogen (385,000 gallons) costs 96 cents a gallon, or $369,000. Total for the two propellants therefore is $423,180, but this does not include the smaller – though nevertheless considerable – quantities of hypergolic propellant used by the Orbiter's OMS (Orbital Manouvering System) and RCS (Reaction Control System). This is a 50/50 mixture totalling 1,700 gallons of monomethyl hydrazine (MMH) and nitrogen tetroxide. MMH costs a staggering $73 per gallon, while nitrogen tetroxide costs $33 per gallon. Total hypergolics bill therefore is $180,200.

Finally, there's 115 gallons of hydrazine, at $84 per gallon, to fuel the Shuttle's auxiliary power units (APUs) and hydraulic power system, which totals-up to $9,660.

Grand total fuel bill for one flight is therefore $613,040 (and average fuel consumption during a mission is an estimated half a mile to the gallon!).

Other costs (quoted here at 'then' rates) make the fuel bill seem very cheap. The vast computer network which orchestrates Space Shuttle operations was programmed by IBM at a cost of $56 million. Launch Pad 39A, which has been used for every Shuttle launch thus far, with the exception of ill-fated 51-L/*Challenger*, was modified from Apollo to Shuttle specification for $44 million.

Pad 39B cost $56 million to convert.

The cost of modifying the Mobile Launch Platforms and Crawler-Transporters to Shuttle specification was $11 million, while the cost of building and outfitting the Shuttle Landing Facility (SLF) at Kennedy Space Center was $27.2 million.

Due to high development costs coupled with a very low production run, the spacesuits worn by Shuttle astronauts when they are working outside the Orbiter vehicle cost an astronomical $2 million *apiece*!

To put all this into perspective, the $10.3 billion total cost of the Apollo program which put Man on the Moon – $88.5 billion at today's prices – allows favorable comparison with $33.4 billion for the same number of Space Shuttle flights.

(This page) **The massive Mobile Launch Platform and Crawler-Transporter vehicles were modified at a cost of $11 million for Space Shuttle operations following years of sterling service in the Apollo era.**

Shuttle Menu

SPACE SHUTTLE FOOD AND BEVERAGE LIST

Foods

Applesauce (T)
Apricots, dried (IM)
Asparagus (R)
Bananas (FD)
Beef almondine (R)
Beef, corned (I) (T)
Beef and gravy (T)
Beef, ground w/pickle sauce (T)
Beef jerky (M)
Beef patty (R)
Beef, slices w/barbeque sauce (T)
Beef steak (I) (T)
Beef stroganoff w/noodles (R)
Bread, seedless rye (I) (NF)
Broccoli au gratin (R)
Breakfast roll (I) (NF)
Candy, Life Savers, assorted flavours (NF)
Cauliflower w/cheese (R)
Cereal, bran flakes (R)
Cereal, cornflakes (R)
Cereal, granola (R)
Cereal, granola w/blueberries (R)
Cereal, granola w/raisins (R)
Cheddar cheese spread (T)
Chicken a la king (T)
Chicken and noodles (R)
Chicken and rice (R)
Chili mac w/beef (R)
Cookies, pecan (NF)
Cookies, shortbread (NF)
Crackers, graham (NF)
Eggs, scrambled (R)
Food bar, almond crunch (NF)
Food bar, chocolate chip (NF)
Food bar, granola (NF)
Food bar, granola/raisin (NF)
Food bar, peanut butter/granola (NF)
Frankfurter (Vienna sausage) (T)
Fruitcake
Fruit cocktail (T)
Green beans, french w/mushrooms (R)
Green beans and broccoli (R)
Ham (I) (T)
Jam/jelly (T)
Macaroni and cheese (R)
Meatballs w/barbeque sauce (T)
Nuts, almonds (NF)
Nuts, cashews (NF)
Nuts, peanuts (NF)
Peach ambrosia (R)
Peaches, dried (IM)
Peaches, (T)
Peanut butter
Pears (FD)
Pears (T)
Peas w/butter sauce (R)
Pineapple, crushed (T)
Pudding, butterscotch (T)
Pudding, chocolate (R) (T)
Pudding, lemon (T)
Pudding, vanilla (R) (T)
Rice pilaf (R)
Salmon (T)
Sausage patty (R)
Shrimp creole (R)
Shrimp cocktail (R)
Soup, cream of mushroom (R)
Spaghetti w/meatless sauce (R)
Strawberries (R)
Tomatoes, stewed (T)
Tuna (T)
Turkey and gravy (T)
Turkey, smoked/sliced (I) (T)
Turkey tetrazzini (R)
Vegetables, mixed Italian (R)

Beverages

Apple drink
Cocoa
Coffee, black
Coffee w/cream
Coffee w/cream and sugar
Coffee w/sugar
Grape drink
Grapefruit drink
Instant breakfast, chocolate
Instant breakfast, strawberry
Instant breakfast, vanilla
Lemonade
Orange drink
Orange-grapefruit drink
Orange-pineapple drink
Strawberry drink
Tea
Tea w/lemon and sugar
Tea w/sugar
Tropical punch

Condiments

Barbeque sauce
Catsup
Mustard
Pepper
Salt
Hot pepper sauce
Mayonnaise

SPACE SHUTTLE TYPICAL MENU

DAY 1	DAY 2	DAY 3	DAY 4
Peaches (T) Beef patty (R) Scrambled eggs (R) Bran flakes (R) Cocoa (B) Orange drink (B)	Applesauce (T) Beef jerky (NF) Granola (R) Breakfast roll (I) (NF) Chocolate instant breakfast (B) Orange-grapefruit drink (B)	Dried peaches (IM) Sausage (R) Scrambled eggs (R) Cornflakes (R) Cocoa (B) Orange-pineapple drink (B)	Dried apricots (IM) Breakfast roll (I) (NF) Granola w/blueberries (R) Vanilla instant breakfast (B) Grapefruit drink (B)
Frankfurters (T) Turkey tetrazzini (R) Bread (2) (I) (NF) Bananas (FD) Almond crunch bar (NF) Apple drink (2) (B)	Corned beef (T) (I) Asparagus (R) Bread (2) (I) (NF) Pears (T) Peanuts (NF) Lemonade (2) (B)	Ham (T) (I) Cheese spread (T) Bread (2) (I) (NF) Green beans and broccoli (R) Crushed pineapple (T) Shortbread cookies (NF) Cashews (NF) Tea w/lemon and sugar (2) (B)	Ground beef w/pickle sauce (T) Noodles and chicken (R) Stewed tomatoes (T) Pears (FD) Almonds (NF) Strawberry drink (B)
Shrimp cocktail (R) Beef steak (T) (I) Rice pilaf (R) Broccoli au gratin (R) Fruit cocktail (T) Butterscotch pudding (T) Grape drink (B)	Beef w/barbeque sauce (T) Cauliflower w/cheese (R) Green beans w/mushrooms (R) Lemon pudding (T) Pecan cookies (NF) Cocoa (B)	Cream of mushroom soup (R) Smoked turkey (T) (I) Mixed Italian vegetables (R) Vanilla pudding (T) (R) Strawberries (R) Tropical punch (B)	Tuna (T) Macaroni and cheese (R) Peas w/butter sauce (R) Peach ambrosia (R) Chocolate pudding (T) (R) Lemonade (B)

Abbreviations in parentheses indicate type of food: T = thermostabilised, I = irradiated, IM = intermediate moisture, FD = freeze dried, R = rehydratable, NF = natural form, and B = beverage.

AT THE CONTROLS

It is the dawn of another long, hot Florida summer day at the Kennedy Space Center. For a place filled with high-tech heavy industry, the site is eerily quiet, minus even a tick from the digital clock, counting seconds in quartz-timed 20th century silence.

In contrast to the flat swamp, in contrast to the gulls that circle as the sun boils from the Atlantic, the Space Shuttle is graceless, ungainly and out of its element.

One can see it from all angles. From one view, it is a machine, polished, processed and ready. In fact, it is a composite of hundreds of noisy small and large machines and turbines, metals, bolts and seals and washers and plumbing.

From another side, it is the fruit of a thousand careers, jutting in white, black and orange from a concrete pad. In the carbon-carbon tip of the nose are the best and most innovative years of far more than a hundred lives; in the curve of the wings are more years and more lives spent calculating, drawing, designing, testing, and designing again.

To another set of eyes, it is a shell, a lifeguard that will protect its inhabitants from the hostilities it must face to spend just a few days as a world unto itself and return home. The machines, materials and careers that compose the spacecraft are a barrier between the constant, often torturous change it will face outside and the safe haven it must maintain within.

(*Above*) Chief Astronaut Dan Brandenstein – a man with a mission.

But from all these angles the world can see, only one person at a time can see this machine truly do its job. The left-hand seat is a single seat, and only an individual can know how it flies, or what it looks like, or what it feels like and sounds like and is like.

He is at the controls and in command, and by the time he takes his seat, he is engrossed in the complexity of the machine and the task it will perform. Dan Brandenstein, Chief of the Astronaut Office, has taken that seat for two flights, and ridden next to it once. Along with the burden of responsibility a Commander carries on board, he carries another when he returns: to describe his journey to the thousands that made it happen. And all they can do is ask.

"When you've had the opportunity to work with the Shuttle and fly in it, you get a real appreciation for the total team," Brandenstein says. "You really see how broad a scope of folks are involved in it and just how complex it is. And then you see it operate as well as it does.

"I really don't think people understand

(Above and left) **Launch is the most dangerous phase of a Shuttle flight. Says Brandenstein: "If you aren't keyed-up when the boosters light, you ought to get keyed-up then''. The Chief Astronaut confides that, from lift-off through the roll program, those on board "don't have much control to fall back on.''**

the versatility of the Shuttle. I don't think they understand what it can do and really how complicated it is. As a crew member, as a Commander, you become more familiar with it. But it's just that the more and the better you know it, the more you understand just how elaborate it really is."

When the silent clock strikes zero, more than 250 NASA television and film cameras are trained on the Shuttle. Commercial television cameras line platforms three miles away, and the scenes on which they focus are transmitted to every major American network. Still-cameras are planted remotely in the swamps around the launch pad; even more cameras click in a press grandstand. And no-one counts those that line the roads for ten and 20 miles around.

At ignition, the solid rockets go from zero to 44 million horsepower in a split second. The main engines produce a power equal to that generated by 23 Hoover Dams.

But all the eyes that study it as a spectacle, both electronic and human, can't see it from the perspective of that single front left seat.

"If you aren't keyed-up, when the boosters light, you ought to get keyed-up then. For a certain amount of time, from lift-off through the roll program, you feel you don't have much control to fall back on. The roll program would be a very difficult maneuver to fly manually. Beyond that, we practice flying ascents manually in the simulator, and I feel very confident that if you had to, you could fly it. But the roll program is a very coordinated manuever. In the simulator you can fly it, but you can't fly it very well and you can't always fly it.

"You know, I've flown three times on the Shuttle, but I've probably looked out the window for only two or three glances on ascent. You are really focused on making sure the vehicle is operating properly. You're cycling through the various displays and you're monitoring trajectory very closely to be sure it's doing what it's supposed to be doing. In that dynamic a region, if it starts to do something wrong, you can't be hesitant. You know what the limits are and when you are going to have to take over manually if something happens."

Eighty NASA cameras still focus on the Shuttle at 150,000 feet, two minutes and ten seconds after launch, when the solid rockets burn out. From the ground, they appear to fall away like arrows at the height of their flight, with a slow grace that camouflages the explosive charges that push them away. The spacecraft is moving more than four times the speed of sound.

"Before my first Shuttle flight, the highest altitude I'd been was probably a little over 50,000 feet.

"When the solids go, you feel a dip. It feels like you're falling for just a second. It doesn't really do anything like that, but it feels that way. After SRB sep is past, it gets so smooth. I always describe it as like a sewing machine. It just kind of purrs.

"My first glance out the windows on a flight was right after the roll program. You look out and then check for debris going by the windows. The sky was still blue. The second glance after that, there was a little bit of blue, but the sky was mostly black."

(Above) **Amply demonstrating the Shuttle's unique capabilities, the 51-A *Discovery* crew retrieved the crippled Westar 6 and Palapa B2 comsats for repair and relaunch.**

For six and a half minutes after the solids have expired, the main engines continue to burn, pulling fuel from the External Tank at a rate that would drain an average swimming pool in less than nine seconds. To those watching, it is simply streaking toward space, continually climbing, constantly accelerating and slowly disappearing. But to those flying, it is passing through boundary after boundary, climbing a set of safety-net stairs to orbit.

"You are always busy monitoring systems. You give the middeck folks a slight play-by-play of what's happening now, what's about to happen, to keep them informed. Your concentration on trying to keep track of options does decrease some, though it is more like it shifts. When you pass the 'Negative-return' call, you don't have to worry about a return-to-launch site abort. That's one less thing you have to worry about, so you kind of flush that. But you still have a variety of abort boundaries to go through.

"Then when you get the 'Press-to-MECO' call (*main engine cutoff – Ed.*), well, your transatlantic aborts are gone, so you kind of relax on that. But you still have more things ahead.

"We're comfortable with the whole system as we have it now. As long as you can override something that's automated if it isn't doing the right thing, and as long as you can do that before it puts you in a situation that you can't recover from, then automating things is better.

"Anytime that you can have more capability by doing it automatically, that's the thing to do. Given computer power and enough sensors, you can automate a lot of things. But anything that happens has to fall within something that you've programmed. You gain by having a man in the loop in a lot of areas, areas where there are so many variables, because he has the ability to take over if something happens that falls out of what the computers can handle. Humans are creative. They're intuitive and can make decisions. You have to take some tradeoffs."

When the main engines cut off, the Shuttle is about 70 miles high and travelling around 17,400 miles per hour. But the sensations in the front left seat say it may as well be sitting still.

"What's really strange is that when the engines cut off, your arms just float up. It doesn't feel like you stopped. The Gs build up the most right before MECO and you're being pushed back, but you aren't thrown forward in your straps when the engines shut down. The acceleration that pushed you in the seat is gone, and you're just floating."

Energy, though it can't be touched or held in a hand, is as real a part of the Universe as a nut, a bolt or a chair. It exists, in one form or another, before it is used and after it is used. Going into orbit is a ballet of energy. The 44 million horsepower per second put out by the solid rockets are still within the Shuttle. The 23 Hoover Dams per second of the main engines have been imparted to the spacecraft. All that remains to put the Shuttle in orbit is a slight boost, an adjustment, a baby step in comparison to a sprinted marathon, without which the spacecraft would descend as quickly as it rose. The Orbiting Maneuvering System engines are a fine-tuning mechanism, easing the Shuttle into a free-fall around the Earth, too fast to come down, too slow to go higher, with a gentle push rarely longer than a couple of minutes.

"Relative to the sensations of ascent, the OMS burn is not much. But it is very noticeable. You're more sensitive. The burn is a bit of a jolt when it starts, and it's just a smooth, gentle acceleration after that. You don't really see yourself get any higher. It is so gradual; you do a burn on one side of the Earth and you don't really see that you're any higher until you're half an orbit away.

"The ground sends information for the burn up to you. The orbit you want is entered in the computer, and it calculates when to do the burn and how long the burn should be. Then you execute the burn, monitor the engines and monitor the trajectory data as the burn continues.

"You don't relax afterward, because then you're into post-insertion. You're configuring for on-orbit operations. You're opening the payload bay doors, you're stowing the seats and getting out of your suits. You're adjusting all the systems for orbit. It's a very busy time and it normally sets the tone for the mission. You want to get through that and stay ahead of the timeline, because as soon as post-insertion is done, then you start getting into the meat of the mission. You don't want to get behind or get caught short.

"It isn't really until the pre-sleep period on the first day when you can really kind of relax and say, 'We're all caught up; We got today done.' Then you can spend more time looking out the window, and you can eat a slow meal. In fact, for lunch on the first day, we usually just carry a bag of sandwiches we can eat on the run."

Brandenstein's first flight, STS-8/*Challenger*, launched at night. The Shuttle reached orbit in darkness, and created its own new day within a few minutes.

"We launched at night, we crossed the Atlantic at night, and, just as we got to Africa, we saw that first sunrise. Of all three flights, seeing that first sunrise is the most memorable to this day. In your training, you get briefed on what things are going to be like: 'Ascent is going to be like this.' But no-one ever said how phenomenal those sunrises were. And it was so gorgeous, it just took my mind away.

"Sunsets and sunrises happen very fast. At sunrise, you see a sliver of sky turn blue, and then you get this tremendous spectra of color all along the horizon. The colors are just so vivid and so bright that it is really amazing."

To the earth-bound, the Shuttle in orbit is at best a twinkle crossing the horizon at dawn or dusk and it is seen then only if conditions, location and timing fall correctly. The strings that bind the ship to Earth are invisible: the gravity that keeps it circling, the signals that bounce off other satellites to a desert dish-antenna and become hundreds of displays and a few voices; a television picture that falls from an object unseen in a clear, blue sky. They would appear tenuous.

"Through your training, you develop this teamwork. Mission Control is there and is another part of the team. You're part of the team. You don't ever feel alone. I have never felt anything like that on-orbit. But you don't feel like they're right next door to you, or like they are always looking over your shoulder, either.

"You have to have what we sometimes refer to in the flying business as 'situational awareness'. You have to have a big picture of all the systems on board. But, then again, by the time we fly, we've been trained by so many experts that we also have a very intimate knowledge of each particular system.

"You have to be able to look at the Shuttle in both ways: an objective view of the whole and a narrow, very focused view of a small part. You move back and forth as the situation dictates. When things are going normally you see the whole. And when you've working a specific problem, you home into the thing at hand. Once you resolve it, you step back to the whole picture. Still, in general, you'll never know everything.

"I sleep on the flightdeck. If something happens during the night, you're right there ready to respond. I sleep in the seat so the rest of the crew can sleep downstairs. If there's an alarm, the controls are right there. You can get to it quickly and not disturb the whole crew, especially if it's something that's not very significant.

"I sleep well on-orbit, but I don't think I sleep as soundly as I do back home. I believe it's mainly because I don't want to get too sound asleep just in case there's

***(Below)* Close encounter. The LDEF free-flyer payload, which Dan Brandenstein's STS-32/*Columbia* crew retrieved after its six-year spell in low-Earth orbit.**

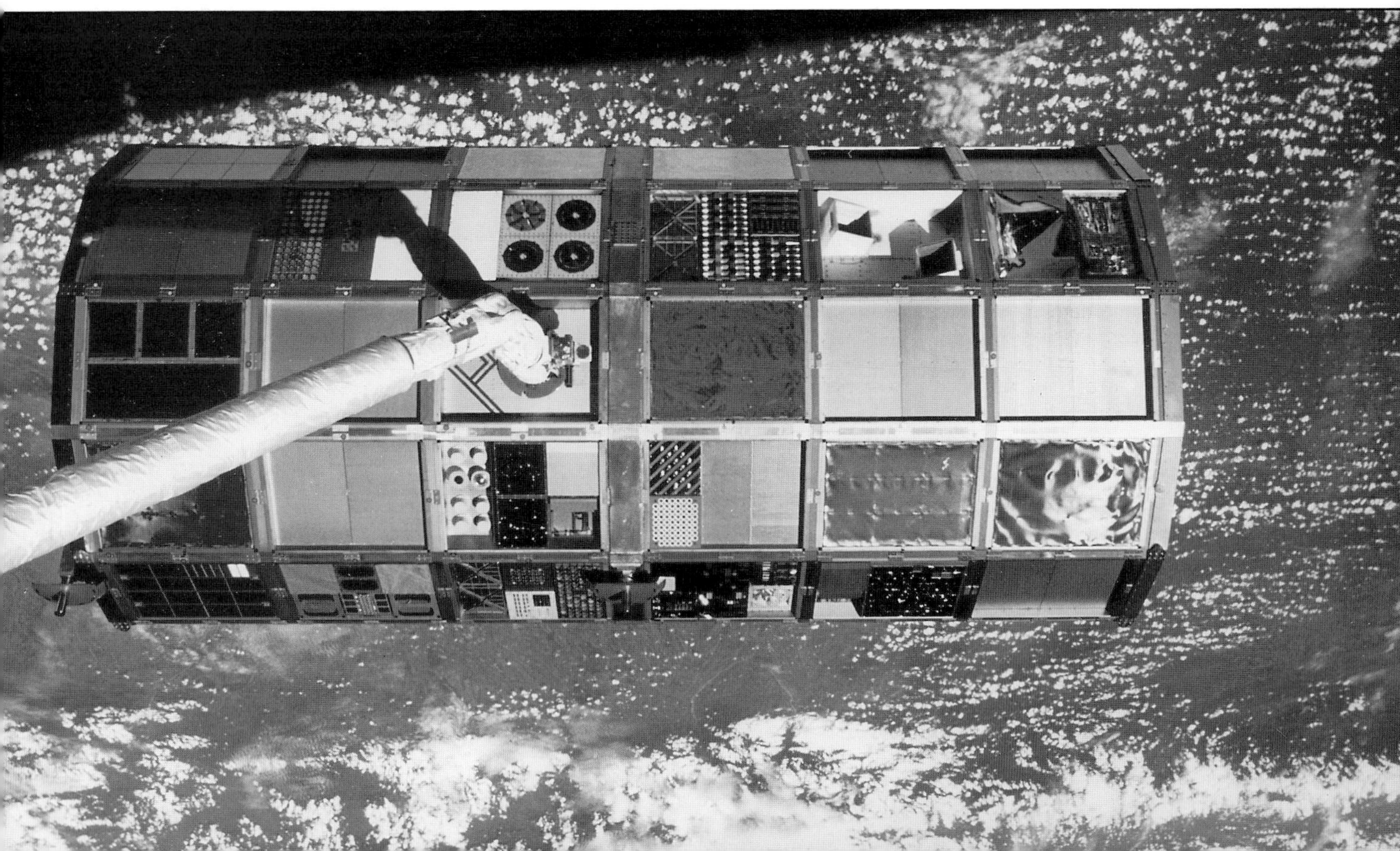

(Above) **The view through the Orbiter's windows is nothing if not spectacular. This photo was taken from the Shuttle's record altitude.**

an alarm or something.

"Day-to-day, when you are doing experiments, eating dinner, or doing housekeeping chores, you let the caution and warning system do its job. You don't monitor things very much. But when you have a spare moment, you go and cycle through the displays."

Flight in orbit is not flight, although it's called that for lack of a better word. Movements of the Shuttle in space are adjustments made to a perpetually falling object through the use of 38 primary jets, six small jets or the two large OMS engines. The wings are simply waiting. Due to the unnatural feel of orbital mechanics, flight is now a precise calculation of cause and effect more than it is a human feel for what will occur and why. The idea of any movement at all is relative to where you are looking.

"You don't have a sensation of speed such as driving fast down the road, because poles aren't whizzing past you, you aren't hearing the rumble of wheels on the ground. It's not even like flying an airplane at low altitude, where you see the terrain zipping past, you feel the turbulence and you hear the wind noise. You're in a silent environment other than the cockpit noise, and the only sensation you have is when you look out the windows and see the ground tracking below you. But to see continents come and go, to take on the order of ten minutes to cross the United States, it's obvious you're really humming. But it's a different sensation of speed. You don't have the acceleration. It's just zero-gravity, floating at almost 18,000 miles an hour.

"We did some preparatory burns the day before we caught up to the Long Duration Exposure Facility and, with those, you don't really feel like you're catching-up to anything. They are just OMS burns. But the night before the rendezvous, when we went to bed, with the sun angle right, we could see LDEF, though we didn't really get close to it until the next day.

"The Reaction Control System feels very tight when you are flying close to another object. You fire the big jets and the vehicle gets a big thump, shakes and it moves. When you get into very close proximity operations, precision flying is based on two things: first, it has a really well-designed flight control system; second, you train a lot.

"You can control it to within inches this way or inches that way relative to the task you're trying to accomplish. We did that when we retrieved LDEF, and we did that when we retrieved Spartan on my second flight. The flight control systems holds an attitude so well, and it makes flying so precise, you can make very specific movements.

"It is a tribute to how well designed it is. That's the type of control you have. Still, you can't always get it perfectly stopped relative to another object, so you might just move it so far and let it take a very slow drift."

The only time the Shuttle's speed is constant is when it is in orbit. To go from zero to almost 18,000 mph in eight and a half minutes is a feat, but the bigger feat is to go from 18,000 mph to zero and remain intact. The Shuttle poses for its return to air and wind and land like the traveler who pulls coat collar tight and puts chin and chest into a bitter winter breeze: it takes a posture of defiance. To designers, crew and flight controllers, it is called the high angle of attack; its nose is angled high, and its most durable portion, the tiles underneath, greet Earth first in a battle of air and speed.

"The deorbit burn feels just like the OMS burn on ascent. But as soon as you're done with it, you pitch around and get the nose forward and up. Then, as you start your fall toward the atmosphere, you do notice that you're coming down. It looks like you are getting closer to the Earth. But you normally go into night very quickly, and then your visual cues are gone.

"Then, in darkness, the first sensation you get is when you are a bit into the atmosphere and the Gs start building up. It still doesn't feel like a descent; it feels like being in an airplane and pulling Gs. It just feels like you're squishing down in the seat. The only real sensation of descent is from watching the altimeter click off.

"From the sensations, without instruments, you wouldn't know the difference between Mach 25 and Mach 1."

The Shuttle's entry is automated from the deorbit burn through three, gradual sweeping S-turns, one of which can take it half an ocean to complete. The atmosphere is the only brake it has to slow it from Mach 25 to 200 miles an hour. The friction between air and spacecraft produces temperatures of almost 3,000 degrees Fahrenheit. To release the energy it received at launch, it creates as much of a spectacle on re-entry.

"You don't get used to seeing the plasma build up. At almost 350,000 feet, you start to see a little pink out of the windows, coming up from the bottom.

"It turns into kind of a pink glow, and, from that, becomes an orange glow. It then becomes a very deep orange, before it turns practically white it is so hot. The plasma flow is that dense. In fact, on the corners of the windows, you can see a turbulent flow with swirls in it. It's sort of like rain on the car window, but it isn't drops, it's a flow pattern.

"Then, at about 180,000 feet, it goes in reverse: the white gets less dense, then it goes to orange, then pink and then it's gone.

"During this phase, you come into daylight. The Earth is still dark, and in those areas you can still see the plasma. But where the sun is rising you can see it, because of the light background. On both extremes, you have the orange, pinkish plasma. In the middle, you have a blue stripe when the sun is coming up. It only lasts a few seconds.

"You've trained to the point where you know, if you had to take over, you could do it. But you don't even hold your hand on the stick all the time during entry. You are just monitoring systems, cycling through displays on the screens, and checking them against your checklist card. It's a very close analogy to an airplane on autopilot, though you are monitoring things very closely.

"When you see the sun come back up again, it's obvious that you're much lower than you had been. But even then, when you break out of night, you don't have a sensation of going down. You still feel mostly just a forward velocity."

As the Shuttle descends and slows, the jets that have kept it stable are replaced by the rudder and elevons for control. Flight becomes flight again in the traditional sense.

"You can hear the primary jets fire, and you can see them in front. But you don't hear wind noise and you don't hear much of anything except what is coming through your headset.

"You don't notice the change from the jets to the aerosurfaces; it's very subtle.

"When we go subsonic, at about 60,000 feet or so, we take over manually, flying it around the Heading Alignment Circle all the way to touchdown. That way, when you get to the landing phase, you're in tune with the vehicle. You're aware of its responses.

"The Shuttle has an autoland capability built in, but it has never been tested. Early in the program, we looked at a possible test of it, but the concern is that if it errs close to the ground, there's nothing you can do to take over. Say it made an error at 50 feet from touchdown, and you've got a pilot who had been in orbit five days, done the entire landing on autopilot and suddenly had to take control. In an unpowered vehicle he'd essentially be helpless. There wouldn't be time to make a proper correction. Trying to take over suddenly without getting used to the vehicle first would be difficult. The tendency would be to overreact.

"On first landings, almost everyone notices a sort of time compression. The events seem to happen faster than they did in the simulator or in the Shuttle Training Aircraft. The STA is very accurate in duplicating the Shuttle, so your landing feels very much the same; you feel like you've done it before. The time compression is probably due to anxiety, because it doesn't seem to happen the second time you land.

"The Shuttle goes through 'Mach buffeting' as it goes subsonic. It's a shaking, kind of like a car going down a gravel road, due to air transitioning from supersonic to subsonic flow over the wings, and it lasts about ten to 15 seconds.

"It flies very crisply. It is very adaptable. The digital flight control system allows it to respond very much the same with different centers of gravity and different weights.

"I've flown 747s and the KC-135, which are big airplanes. In them, you have a certain lag in the responses. But the Shuttle flies more like a fighter than a big airplane. You know you are flying a large aircraft, but the controls are positive and crisp.

"You don't get much of a sensation of descent until you drop the nose on final. The approach pattern is much different than a fighter, carrier approach or anything else. In the Shuttle you have no power, and most of the time you're constantly decelerating. On the outer glideslope to the runway, maintaining a constant speed of 290 knots, – and it's pretty steep – you're kind of hanging in the straps then. You keep the speed constant by opening and closing the speed-brake. You feel the speed-brake take hold and you feel the drag in general."

On its final approach to the runway, the Shuttle descends seven times more steeply than a commercial airliner. It is dropping from the sky 20 times as fast. Less than 2,000 feet above the runway, it pulls up to reduce its angle of descent to just slightly less than that of an airliner. Its final maneuver before touchdown is a slight flare upward of the nose, to slow it even more and allow a gentle easing down of the nose landing-gear after the main gears have touched Earth.

"You don't feel the final flare. The only big difference on touchdown is between the lakebed and the concrete runway. Rollout on the runway is much smoother. The lakebed is pretty rough.

"If you bring it to touchdown right, you hardly notice it. It's smoother than a landing in a commercial jet.

"When it was first being designed, they said the Shuttle was going to fly every two weeks, 60 missions a year. That was obviously far too optimistic. If you could simplify the turnaround, reduce the care and feeding, then that would be a big help. But I don't consider today's flight rate to be any type of a drawback on the vehicle, simply because there's nothing else like the Shuttle, nothing else that can do what it can do. And there never has been.

"It is being part of a team that accomplishes a mission that you remember. That's the whole thing. You take it in steps. You're proud of every part of it; you're proud of your crew; you're proud of everyone that worked on the flight; and you're proud of the whole team that made the Shuttle perform. When you walk away and you're all done, that's what you remember and that's what really makes you feel good."

(Below) **"'If you bring it to touchdown right, you hardly notice it," says Dan Brandenstein.**

INTERVIEW: ASTRONAUT TAMMY JERNIGAN

Dan Brandenstein's vivid description of flying the Space Shuttle, gracing the previous few pages, is a pilot's view of events: no more, no less. There's another dimension to his story – that witnessed by the Mission Specialists, who actually constitute a majority within NASA's astronaut corps. As with the pilot-astronauts, groups of Mission Specialists are recruited at biennial intervals. It follows that, at any given time, some of these astronauts are seasoned veterans nearing the end of their spacefaring days, while others are comparative fledglings, their careers stretching far before them.

When a new group of recruits join the astronaut corps, there are always a small number of their predecessors still awaiting their maiden space missions. One such person was Tamara Jernigan, who, at the time this edition was being prepared for publication, was in the final, intensive phases of training prior to entering space for the first time. The opportunity to interview an astronaut so close to flight provided a fascinating glimpse into a world all too few of us get to see.

Tammy Jernigan was born on 7 May 1959 in Chattanooga, Tennessee, but grew up in southern California. She has long brown hair and blue eyes, is five feet six inches tall and weighs 125 pounds. After graduating from Santa Fe High School in Santa Fe Springs in 1977, she entered Stanford University, receiving a bachelor of science degree in Physics (with honors) in 1981 and a master of science degree in Engineering Science in 1983.

After graduating from Stanford, Tammy worked in the Theoretical Studies Branch at NASA's Ames Research Center near San Francisco, California, from June 1981 until July 1985. Her research activities were devoted to astrophysics, with particular emphasis on the study of bipolar outflows in regions of star formation and gamma-ray bursters. During this period, she continued her studies at Stanford, and later the University of California at Berkeley, emerging with a master of science degree in Astronomy from the latter in 1985. She went on to pursue a PhD in Astrophysics at the Department of Space Physics and Astronomy at Rice University, investigating shock-wave phenomena in the interstellar medium.

Tammy was selected as an astronaut candidate by NASA in June 1985. In July 1986, she completed the mandatory one-year training and evaluation program, qualifying her for assignment as a Mission Specialist on future Shuttle flightcrews. That assignment came when she was named to the STS40/*Columbia* crew for the SLS-1 life-sciences-dedicated mission of summer 1991.

Away from the professional sphere, Tammy Jernigan fills her hours with sports. As an undergraduate, she competed in intercollegiate athletics on Stanford's varsity volleyball team. Today, she enjoys volleyball, racquetball, tennis, softball and flying.

***(Above)* Tammy takes a break for refreshments prior to undertaking a spacewalk simulation in the WET-F neutral-buoyancy tank at Houston.**

MACKNIGHT: How did you become an astronaut: did you make a specific beeline for that as a career, or did you just somehow drift into it?

JERNIGAN: When I was growing up, I was always interested in the space program and I liked science and I liked the idea of flying someday – getting a Private Pilot's License, for instance – so when I found out, when I was an undergraduate at Stanford, that NASA had started taking scientists *en masse* from the scientific community (*in 1978 – Ed.*), I began to feel that at some point I might have a chance at getting selected and decided to apply at a later date.

MACKNIGHT: Was it a surprise to be asked to come forward to the interview stage, or are you the sort of person who felt confident that you'd at least get to that level of selection?

JERNIGAN: I'd say I was delightfully surprised. I was a graduate student at the time: I was getting a PHd in astronomy and astrophysics at Berkeley, so it was a pleasant surprise to be interviewed as a graduate student. I was also affiliated with Ames Research Center, working in the space science division, so I had been working for NASA since I was 19.

MACKNIGHT: You were taken on as an astronaut-candidate at the age of only, what, 26?

JERNIGAN: That's correct. I was interviewed when I was 25, and then had a birthday before I started work here at JSC.

MACKNIGHT: So are you the youngest member of the astronaut corps?

JERNIGAN: No longer. I was up until the new class came in. There is now a fella (*Leroy Chiao*) who's about a year younger than I am, who was selected in 1990. He would be the youngest person right now in the Astronaut Office.

I'm happy to pass that baton!

MACKNIGHT: When you were first attracted to the Shuttle program, there were plans for it that haven't all come to pass – the *Challenger* accident, I suppose, was a turning-point in the transition from adherence to theoretical goals to the widespread recognition of the constraints of practical realities. Is there a feeling of disappointment, among the newer astronauts in general, but on your part in particular, that the Shuttle isn't quite able to do everything in the way it was originally intended?

JERNIGAN: Well, I'll just speak for myself. I was certainly disappointed when the *Challenger* accident occurred, not just due to the setback in my own career, but also the loss of human life and the setback to the program. But it seems that the road into space is a rocky one and we're trying to do very ambitious things, so there are going to be some setbacks and it's part of the risk you take when you set out to do things like go into space or try to embark upon a Space Station project.

So, of course, I would like to do everything right now, as quickly as possible, but I understand that can't always be the case, so let's keep making progress and I'll continue to be interested in being part of a space program like we have in the United States, that is ambitious and tries to do new things. We'll begin construction for Space Station soon, I hope, and I'd certainly like to be a part of that.

MACKNIGHT: That's leads neatly on to my next question. You're, effectively, at the very start of your career as an astronaut, at least

(*Above*) Greg Harbaugh is a relative newcomer to NASA's astronaut corps. He is pictured being lowered into the WET-F facility for water survival-training during preparations for flight.

(*Above*) Test pilot Eileen Collins is set to take the assimilation of women into the astronaut corps one step further. In July 1991, she began the long process of training and experience-gaining that will make her the world's first female spacecraft Commander.

***(Above)* All astronauts undergo wilderness survival-training. Mae Jemison seems to be coping well in deepest Washington state.**

in terms of now having the ability to go out and do the things you've trained to do. Notwithstanding the fact that's it's difficult to project ahead, and that you're almost certain to be around for the start of the Space Station *Freedom* venture, do you see your time as an astronaut as a truly career-long endeavor, or could it be the means to some still-greater end somewhere down the line?

JERNIGAN: I think that there's a good chance that I'll be in the space program for many years, because I believe in the types of things they're doing. I'd like to not only be part of constructing Space Station, but perhaps do a tour on the Station once we get it fully up and running.

I envision a long and productive career with NASA.

MACKNIGHT: As far as you've seen so far, what's the best part of being a NASA astronaut?

JERNIGAN: I think it's the diversity the job offers. You get to fly, you get to work in the spacesuit and the water-tank, spend time in the simulators, you get to do some science – really the exposure to all different types of science and engineering and operations has made the job very interesting.

MACKNIGHT: And the worst part?

JERNIGAN: The worst part, perhaps, is things not happening as quickly as you might like them to. You know, one is always anxious to be part of all these exciting things *right now,* and so sometimes it's a little difficult to have to wait – but it's part of the space business.

MACKNIGHT: If you look at the time some people have had to wait to fly, you've been one of the lucky ones. Take Bruce McCandless as an extreme case: after joining NASA as an astronaut, he waited 18 years to get his first flight in space – and Don Lind waited 19 years!

JERNIGAN: Timing is important!

MACKNIGHT: About you're T-38 flying. You said earlier that you liked the idea of flying: did you actually get your Private Pilot's License before you joined NASA?

JERNIGAN: I had a solo license. That's a license you earn on your road to getting your Private Pilot's License: it means you can fly by yourself, but you can't take a passenger.

MACKNIGHT: So what about your T-38 flying? You're either going to love it or hate it. What are you feelings toward being subjected to the rigors of high performance jet flight?

JERNIGAN: Oh, I love it. It's great. As a civilian, it's tough to pay for jet time! It's really been a privilege to be able to learn about flying a jet. We have instructor pilots who are very good at teaching us how to fly the airplane, and then also a lot of the guys that we fly with are generous in terms of giving the back-seaters stick time.

MACKNIGHT: How do you feel about the prospect now of actually flying on the Shuttle? The *Challenger* accident brought home to everybody just how dangerous it can be. Is it a big part of getting ready for the flight – weighing the risk element – or is that very much towards the back of the queue in terms of what you're thinking at this point?

JERNIGAN: (*laughing*) I don't even know that it's in the queue! I'm so busy getting ready for flight I'm mostly concerned that I have a good knowledge of all the Orbiter systems, so that if there is a problem, that we know what the right thing to do is. And also making sure that I'm familiar with all my experiments, so that when we get on-orbit we'll get the most science from the time we spend in space.

MACKNIGHT: So at what point do you weigh the risk: do you think you'll weigh it closer in to flight, or have you already weighed it very early on and have now put it out of the picture to enable you to concentrate on the task in hand?

JERNIGAN: I thought about it when I joined the program, but I feel confident – especially after the *Challenger* accident – that we're really attuned to safety at NASA and that we're gonna do the right thing and make the right decisions. Several Orbiters have been rolled back from the pad because it was felt that they were not ready to fly, or there were some concerns about them flying – some of the problems they discovered late into the flow – so I think that demonstrates NASA's willing to hold off on an aggressive flight-rate if safety is being compromised.

Once you decide to fly on the Shuttle, there really isn't a great deal to be gained by spending much time pondering the risks. You're better off spending your time getting ready to fly.

MACKNIGHT: I guess at this point you must be totally involved in the training and have no time for anything else. Can you tell me a little bit about the training you've been doing for this mission, as opposed to the initial basic training you did when you first joined NASA?

JERNIGAN: We do a lot of flight-specific training, so we've spent a lot of time practising the experimental procedures for a nine-day microgravity mission doing biomedical research. So I've learned a lot about cardiovascular and cardiopulmonary physiology and cell biology. That's been a lot of fun for me, because my background is in astronomy and astrophysics, so I've learned something about an area that I really didn't know much about.

Also, I'm the Flight Engineer on the flight, so the Pilot, Commander and I spend a lot of time in the simulators, learning how to troubleshoot malfunctions – in perhaps a computer system or electrical system or environmental control system – so that we know immediately what the right thing to do is.

I've spent a lot of time in simulators!

MACKNIGHT: Have you gone on travel to explain this to a wider audience, or will you do that more after the flight?

JERNIGAN: Both. After the flight, we'll certainly spend a fair amount of time doing that. Prior to going into the intensive phase of training for this mission, I've gone to schools, I've spoken at conferences – it kind of runs the gamut, from meetings of girl-scouts and a governors' symposium, to visits to various companies that build parts for the Shuttle. There are many different audiences, and the one thing they all have in common is just a great interest in the space program.

MACKNIGHT: Can you tell us a little more about your upcoming SLS-1 flight?

JERNIGAN: The main objective of our mission is to get a better understanding of how the human body adapts to weightlessness and then readapts to the one-G field upon return to Earth. We're trying to do a lot of quantitative work on the cardiovascular system and the lungs and the blood. Your body is seriously effected when you go into a microgravity environment, and if we're talking about putting people into space for long periods of time, it's very important that we understand exactly how your body reacts to space and then also, when you come back to Earth, how it has to readapt to the Earth's gravity.

I think that our mission objective is really critical to the future of manned spaceflight.

MACKNIGHT: Particularly to the long-term flights.

JERNIGAN: You bet. There are also applications to the well-being of people here on Earth. One of our investigations looks into the immune system and how the immune system fights infection, and that is directly applicable to all of us enjoying a healthier life here on Earth.

MACKNIGHT: Tammy, thanks for making time to talk – and good luck with your mission.

(Above) **A picture that illustrates far better than words ever could just how diverse Shuttle crews have become. These are the STS-40/*Columbia* astronauts, led by veteran Bryan O'Connor, second from left.**

SHUTTLE AS A STEPPING-STONE

As well as facilitating a host of different tasks in low-Earth orbit, the Shuttle is employed as a 'stepping-stone' to much higher Earth orbits – and, indeed, to other planets in our Solar System. Use of two different auxiliary booster units, McDonnell Douglas's Payload Assist Module (PAM) and the more powerful, Boeing-built Inertial Upper Stage (IUS), has resulted in many spacecraft being placed in geosynchronous orbit, 22,250 miles above the equator, including NASA's own TDRS 'constellation' of tracking and data-relay satellites.

What particularly interests us in this short chapter, however, are the spacecraft the Shuttle has deployed to far more distant destinations: to Venus, Jupiter and its moons, the asteroids, and the Sun itself.

Getting planetary spacecraft into low-Earth orbit, then setting them on their way was always a key objective of the Shuttle program. The first planetary mission launched by the Shuttle was Magellan, designed to radar-map the surface of Venus through its thick cloud-cover. Magellan was set free from *Atlantis's* payload bay in May 1989, during the STS-30 mission. Magellan's cruise through interplanetary space lasted 15 months. The spacecraft was largely assembled (by the Martin Marietta company) from spare parts left over from the Viking, Voyager, Galileo and Ulysses programs.

As these words are written, Magellan's powerful synthetic-aperture radar system is successfully penetrating Venus's sulfuric-acid-laden atmosphere to reveal details of its intriguing surface features. Orbiting Venus every three hours and nine minutes, Magellan was designed to complete its first radar-mapping cycle in 243 days. During that cycle, it acquired images, radiometry and altimetry, and gravitational data from 70 to 90 percent of the planet's surface. If all goes according to plan, six, seven, eight or even nine such cycles will be completed.

In a best-case situation, the entire surface area of Venus can be mapped. In addition, features of particular interest could be 'spotlighted' – i.e. viewed from a variety of angles to provide more detailed information. The scarce commodity that will determine when Magellan's mission must end is the propellant for its attitude-control thrusters. When this is exhausted, Magellan will not be able to point its sophisticated suite of science instruments at their targets.

Also deployed from the Space Shuttle, the U.S./West German Galileo spacecraft is the centerpiece of the most ambitious planetary mission ever attempted. At $1.3 billion, the Jupiter-explorer is certainly the most expensive. Galileo was originally scheduled for launch aboard the Shuttle *Atlantis* on mission 61-G in May 1986, but the *Challenger* disaster put paid to that plan, and the gold, black and silver machine remained grounded for several years – with upkeep overheads of a cool $5 million per month.

Galileo was finally launched by the STS-34/*Atlantis* crew in October 1989. Once it was gently sprung free of its special cradle within the Shuttle's payload bay, its IUS upper-stage boosted it from low-Earth orbit into interplanetary space. It is currently getting a series of gravitational 'kicks' from close-range flybys of Venus (in February 1990) and the Earth (December 1990 and December 1991), increasing its energy prior to departing for Jupiter. It will also speed through the Asteroid Belt twice, encountering the asteroids Gaspra (in October '91) and Ida (in August 1993).

Galileo will explore the giant red planet Jupiter in unprecedented detail. It will gather information in three distinct areas of Jupiter's environment: the upper levels of the hostile atmosphere surrounding the planet; the magnetosphere – or magnetic

(*Below*) The first planetary spacecraft launched from the Shuttle was the Magellan Venus radar-mapper, in May 1989.

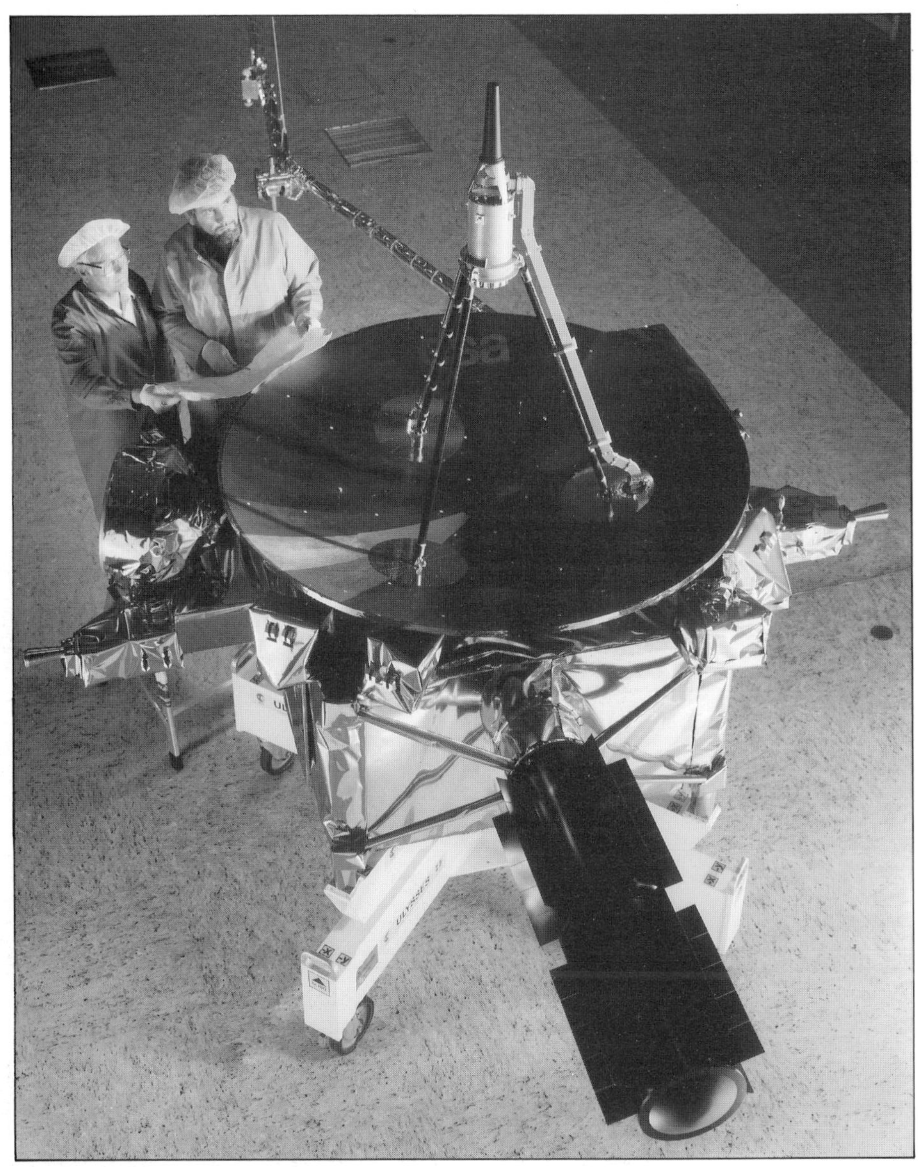

***(Above)* Europe's Ulysses solar explorer was finally launched in October 1990, after a series of frustrating delays.**

field – around the planet; and the Galilean satellites, Jupiter's complex system of moons, with a view to determining their chemical composition.

To accomplish this unique and very demanding mission, Galileo will transform itself from one spacecraft into two (Orbiter and Probe) some 150 days before arrival at Jupiter. The Orbiter element will release the Probe and then deflect its trajectory to avoid a collision with the giant planet, aiming instead to enter orbit around it. As it orbits, the Probe element will head straight for Jupiter. Its task: to enter the very atmosphere of the inhospitable gas-planet and relay information back to the Orbiter, which by this time will be oriented to receive those signals and direct them, via mission controllers at the Jet Propulsion Laboratory in Pasadena, California, to scores of scientists back on Earth.

The Probe does not have the ability to maneuver, so the Orbiter's speed and positioning at the time of the Probe's release must be absolutely accurate to ensure the latter's entry into Jupiter's atmosphere at a flightpath angle of 7-10 degrees. At shallower angles, the Probe would just ricochet off the top of the atmosphere, skipping back into infinite space. At steeper angles, on the other hand, fearsome aerodynamic forces would simply crush the tiny capsule...

It's a one-way trip for the Galileo Probe. Its journey down into the atmosphere, suspended beneath its own heat-resistant parachute, will last no more than 60 minutes until the steady built-up of heat and pressure (the latter, 20 times that of Earth's) destroy it. As soon as this destruction takes place, the Orbiter element – flying high overhead – will become a ship of discovery in its own right. For 22 months it will circle Jupiter, making numerous close-range flybys of its four moons (Io, Callisto, Ganymeade and Europa) and studying its magnetic field.

Like Galileo, the European Space Agency's Ulysses spacecraft had to sit and wait its turn to ride the Shuttle. Had *Challenger* not exploded, it would – again like Galileo – have been launched in May 1986. It was finally set on its way during the STS-41/*Discovery* mission in October 1990. The objective of the Ulysses mission is to explore the Sun's environment (rather than the Sun itself) from a unique vantage-point. That vantage-point is a solar-polar orbit – an orbit that will take the instrument-laden craft over the North and South poles of the Sun for the very first time.

Up until now, we have seen the Sun as though it were projected onto a screen in just two dimensions, but Ulysses will allow scientists to study the Sun in a third dimension. Continuous observations from this unique perspective – far above the plane of the Earth's orbit: the ecliptic plane – will provide an enormously improved understanding of the Sun's behavior, which may in turn help us to predict such things as changes in the Earth's climactic conditions and reap a wide variety of other benefits.

To get into an orbit which crosses the poles of the Sun, Ulysses had to be shot up out of the ecliptic. No rocket engine has the power to accomplish this on its own, so Ulysses employed the gravitational pull of the planet Jupiter to act as a slingshot and bend its trajectory away from the plane of the ecliptic and set it on its way towards the poles of the Sun.

Built under the prime contractorship of Dornier Systems of Friedrichshafen in Germany, leading the pan-European STAR industrial consortium, Ulysses is the sole survivor of what was originally conceived as a two-spacecraft program known as the International Solar Polar Mission (ISPM). Approval was forthcoming in 1978/9 for two craft – one to be supplied by ESA, the other by NASA – to fly a complementary mission. However, a NASA budgetary cutback led to the cancellation of the American element in 1981. The U.S. craft was to have carried cameras to photograph the Sun's surface and inner atmosphere (the solar corona) from the unique vantage-point of the trans-polar orbit.

Those potentially starling views of the Sun are lost to us now, but despite the American withdrawal the program has remained a joint ESA/NASA venture, to the extent that the European craft continues to carry a mixed U.S./European scientific payload. In addition, America furnished the necessary launch services (Shuttle/upper stage) at no expense, together with the RTG nuclear power source and tracking facilities of the duration of the mission.

Although the *Challenger* disaster can be blamed for the fact that Ulysses was delayed several years, the ESA spacecraft – which was essentially completed way back in late-1983 – was originally scheduled for a Shuttle launch on a date as early as February 1983! Problems at the American end associated with developing a suitable upper-stage capable of carrying Ulysses from the Shuttle's low-Earth orbit into an interplanetary trajectory meant that this launch window, and a similar opportunity in April 1985, were missed.

During the development life of Ulysses, the upper-stage earmarked by NASA to boost it on its way was frustratingly changed time and again: first from an IUS to a Shuttle/Centaur, then back to an IUS – which has a lower weightlifting capability, forcing design changes on Ulysses – then back to a Shuttle/Centaur again. Soon after the *Challenger* accident, the Shuttle/Centaur was cancelled altogether on safety grounds. Ulysses was back on an IUS!

A VIEW OF THE FUTURE

(*Above*) Deployed by the Shuttle *Discovery* in April 1990, the Hubble Space Telescope (HST) was soon found to suffer from a misshapen primary mirror. A Shuttle rescue mission is now being planned for 1993. Spacewalking astronauts will fit a device analogous to spectacles to correct the telescope's vision.

Several exciting events detailed in this section of the previous edition, *Shuttle 2*, have now taken place: deployment of the Magellan, Galileo and Ulysses interplanetary spacecraft and orbiting of the Hubble Space Telescope. Two other programs have not fared so well. The OMV 'space tug' became a casualty of NASA's new-found awareness of financial/political realities. Had it come into being, the OMV, or Orbital Maneuvering Vehicle, would have been regularly carried into orbit by the Shuttle. It would have been capable of ferrying payloads from place to place under the command of pilots back on Earth, climbing to as high as 1,500 kilometers – well above the operating range of the Space Shuttle.

Another casualty of the harsh political climate has been the Space Station *Freedom*, which, as this edition closes for press, stands a 50-50 chance of being cancelled altogether. If it survives a forthcoming series of battles on Capitol Hill, *Freedom* will be orbited in component form by the Shuttle and assembled by robots and spacewalking astronauts, fulfilling – at least in part – the dreams of those who see this as America's 'Next Logical Step in Space'.

The fabulous Hubble Space Telescope (HST), orbited in April 1990, has posed some significant problems for those tasked with extracting the maximum scientific benefit from it. This massive spaceborne observatory was named in honor of the American astronomer Edwin P. Hubble, who refined the Belgian abbot Georges Lemaitre's theories relating to the 'Big Bang' theory of an ever-expanding Universe – something HST will itself investigate. Having already been long-delayed by the *Challenger* accident, it has been subsequently found to suffer from a misshapen primary mirror: the result of an infinitesimal calculating error during the early stages of its manufacture a decade earlier.

A Shuttle rescue mission is now being planned for 1993: an option made possible by a happy combination of the Shuttle's unique capabilities and HST's 'spacewalker-friendly' design – it is the first spacecraft designed specifically to be serviced on-orbit. When the spacewalkers reach HST, they'll fit a device analogous to spectacles to correct the telescope's vision. In the meantime, astronomers are using computer processing techniques to enhance the images HST returns, and many significant discoveries have already been made.

Over the next few years, NASA will gradually increase the duration of Shuttle missions, from ten days to 16. One day, 28-day missions may even be possible. Longer stays in orbit will come about once the first of the EDO – Extended Duration Orbiter – spacecraft come into service. In the summer of 1991, *Columbia* was returned to Rockwell's Palmdale, California facility for conversion to EDO standard, and

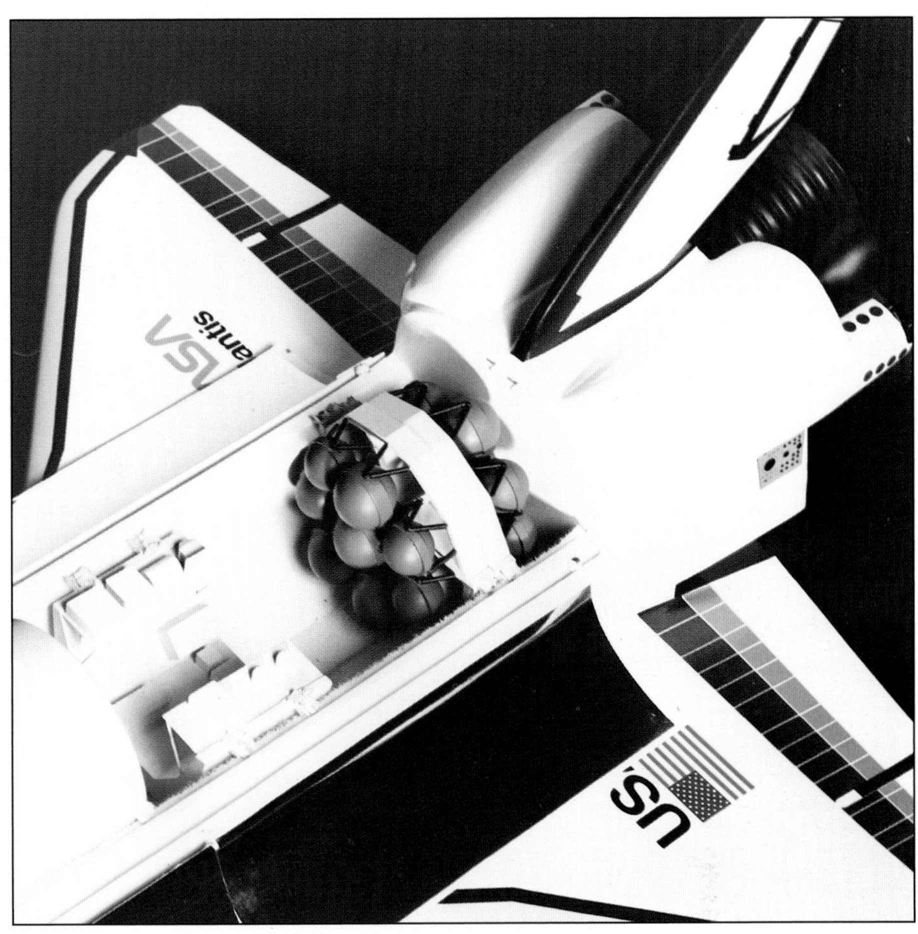

***(Above)* To facilitate extended-duration Shuttle missions, this pallet of additional 'cryosets' will be installed in the Orbiter's aft payload bay.**

***(Below)* A full-scale mockup of the proposed HL-20/PLS mini-shuttle at NASA's Langley, Virginia facility.**

the new Orbiter, *Endeavour,* was equipped to this specification from the outset.

The major addition to the Orbiter for extended-duration missions is additional cryogenics to feed the fuel-cells that generate electrical power for the spacecraft's systems. *Columbia* presently is equipped with five 'cryosets' of liquid hydrogen and oxygen tanks. Each set provides about two days of electricity to the vehicle, using an average of 18 kilowatts of power. A pallet designed by Rockwell and mounted at the rear of the Orbiter's payload bay has the capability of carrying an additional four sets. A series of plumbing lines joins the honeycomb-shaped pallet, which weighs about 6,450 pounds loaded, with the other tanks to form an integrated set.

Improving the process of storing solid waste will also be necessary for longer flights, due to the storage limitations of the current Waste Containment System (WCS), or space toilet. The current WCS has to be removed after each flight for cleaning. It has a limited capability and won't hold the waste of seven crewmembers for 16 days. The improved potty collects the waste in bags that are then compacted into a storage tube. When the tube is full, a charcoal filter, along with odor and bacteria filters, are placed on top. The tube is stored in a locker and another tube is placed in the WCS. After the Orbiter lands, there'll be no need to take out the potty and clean it. It will only be necessary to take out the tube and dispose of it.

Implementation of these and various other modifications will permit the first EDO mission to take place in mid-1992. EDO missions will feature the European-developed Spacelab manned module, or possibly the smaller Spacehab module, due to make its flight debut in 1993.

At the time of writing, plans to employ existing pieces of Shuttle hardware to create a new-generation heavy-lift launch vehicle were continuing to solidify. The concept, designated Shuttle-C (for cargo), would comprize a Shuttle External Tank and two solid rocket boosters mated to a new cargo-canister mounted where the Orbiter vehicle would normally be. Shuttle-C would be capable of hefting much heavier payloads into orbit than the 'conventional' manned vehicle, and would suit those occasions when it is not necessary to have humans aboard to achieve mission objectives. By utilizing existing hardware where possible, Shuttle-C development costs could be kept comparatively low.

No account of the possibilities for the Shuttle program's future would be complete without a brief description of the reusable mini-shuttle concept currently under study by personnel at NASA's Langley, Virginia facility, working in conjunction with Rockwell International and two North Carolina universities. Designated HL-20/PLS – the initials stand for Personal Launch System – this vehicle is effectively a scaled-up version of the Northrop HL-10 manned lifting body of the 1960s, envisioned as a low-cost alternate means of getting people and small but valuable items of cargo to and

***(Below)* An artist's impression of Space Station *Freedom* in 1999, in its permanently-manned configuration.**

from low-Earth orbit should the Space Shuttle be unavailable. It could complement the existing Shuttle vehicle and would enhance operational flexibility when Space Station *Freedom* becomes fully operational.

According to present plans, the HL-20/PLS would be launched from Kennedy Space Center atop a new 'man-rated' version of the Martin Marietta Titan 4 expendable booster. It would measure about 29 feet from nose to tail, have a wingspan of about 24 feet, and weigh about 22,000 pounds. By comparison, the current Shuttle Orbiter weighs around 185,000 pounds. Unlike the present Shuttle, it would provide its crew with a means of escape during the hazardous launch/ascent phase. For a typical Space Station crew-transfer mission, the little 'space taxi' would carry up to eight *Freedom* personnel and two flight crew members.

With its wings folded, the HL-20/PLS could fit in the Shuttle's payload bay. Alternatively, it could be permanently docked with the Space Station *Freedom* for use in dire circumstances, such as when a crewmember becomes critically ill and needs to be returned rapidly to Earth, or if the crew must abandon the station because of a fire or other on-board emergency.

One of the keys to successful, low-cost use of the HL-20/PLS will be its optimization for what's termed 'maintainability'. By emphasizing design simplicity – existing Shuttle-program technologies, such as thermal-protection materials, would be widely employed – and adopting aircraft-type maintenance practices, it is estimated that the HL-20/PLS will require only about ten percent of the processing time required by the current Shuttle Orbiter between missions.

In the much longer term, we are bound to see a new generation of single-stage-to-orbit vehicles entering the arena. One particularly radical proposal has emanated from the studios of the flamboyant German-born industrial designer Luigi Colani. His 'Space Shuttle Polymorph' makes the present Shuttle look positively pedestrian!

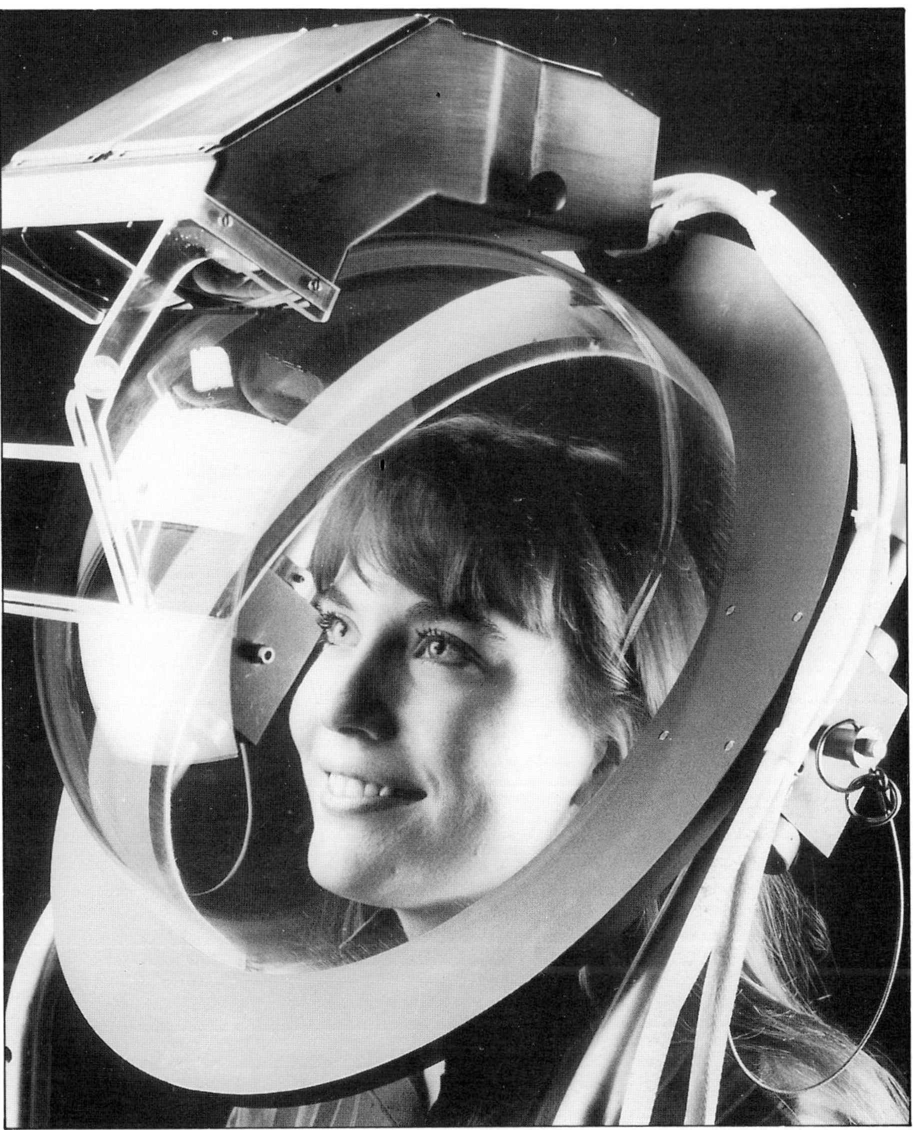

(Above) **New items of hardware are being developed continually. In scaled-down form, this prototype helmet-mounted computer monitor might find its way into NASA spacesuits.**

(Below) **German industrial designer Luigi Colani has proposed that this fabulous 'Space Shuttle Polymorph' replace today's Shuttle.**